Lecture Notes in Computer Science   13822

More information about this series at https://link.springer.com/bookseries/558

Maria De Marsico · Gabriella Sanniti di Baja ·
Ana Fred (Eds.)

Pattern Recognition Applications and Methods

10th International Conference, ICPRAM 2021
and 11th International Conference, ICPRAM 2022
Virtual Event, February 4–6, 2021 and February 3–5, 2022
Revised Selected Papers

Springer

Editors
Maria De Marsico
Sapienza Università di Roma
Rome, Italy

Gabriella Sanniti di Baja
ICAR
Consiglio Nazionale delle Ricerche
Naples, Napoli, Italy

Ana Fred
IST - Torre Norte
Instituto de Telecomunicações
Lisbon, Portugal

University of Lisbon
Lisbon, Portugal

ISSN 0302-9743   ISSN 1611-3349 (electronic)
Lecture Notes in Computer Science
ISBN 978-3-031-24537-4   ISBN 978-3-031-24538-1 (eBook)
https://doi.org/10.1007/978-3-031-24538-1

This Springer imprint is published by the registered company Springer Nature Switzerland AG
The registered company address is: Gewerbestrasse 11, 6330 Cham, Switzerland

Preface

The present book includes eight papers that are the extended and revised versions of a group of selected papers from the 10th and 11th editions of the International Conference on Pattern Recognition Applications and Methods (ICPRAM 2021 and ICPRAM 2022).

Since its first edition, the International Conference on Pattern Recognition Applications and Methods (ICPRAM) has aimed to be a major point of contact between researchers, engineers, and practitioners in the areas of pattern recognition and machine learning, both from theoretical and application perspectives. Therefore, the topics of the conference papers span a wide range of investigation as well as development lines, which of course always reflect the recent trends of research in the pattern recognition community.

ICPRAM 2021 received 97 paper submissions from 30 countries, out of which 3% were selected. ICPRAM 2022 received 107 paper submissions from 33 countries, out of which 5% were selected. The selection was made by the event chairs and their choice was based on a number of criteria that include the evaluations and comments provided by the Program Committee members and by the session chairs during the event. The authors of the selected papers were then invited to contribute to this book by submitting a revised and extended version of their conference papers having at least 30% innovative and relevant material. All the papers clearly reference the conference work and also underline the nature and content of the extension. In particular, the first three papers in the book are the updated versions of contributions presented at ICPRAM 2021, while the following five papers are the updated versions of contributions presented at ICPRAM 2022. These papers contribute to the research on several different topics tackled during the event, including but not limited to deep learning and neural networks, image and video analysis and understanding, machine learning methods, model representation and selection, knowledge acquisition and representation, feature selection and extraction, ensemble methods, data mining and algorithms for big data, biometrics and bioinformatics, and systems biology.

The ICPRAM 2021 paper "Similiarity Constrained Conditional Generative Autoencoder with Generalized Dilated Networks" deals with the popular topic of generative adversarial networks (GANs). Disentangled representations are considered to be an important component for the property of interpretability, which may help overcoming some limitations in the plausibility of GAN-generated results.

The next paper from ICPRAM 2021, "Forecasting Overtime Budgets for Naval Fleet Maintenance Facilities Using Time-Series Analysis During Transient System States", discusses the amendments necessary to adapt the predictive models that were initially developed to predict an annual budget for staffing overtime hours within a Royal Canadian Navy (RCN) fleet maintenance facility. In particular, the drastically changed working conditions caused by COVID-19 revealed some model flaws that this work aims at fixing by adopting a novel strategy.

The last paper from ICPRAM 2021 is "Reduced Precision Research of a GAN Image Generation Use-case" which again deals with GANs, but this time with a specific use-case. The paper addresses a concrete problem of complexity reduction of GAN-generated models. In particular, a GAN model is quantized after training to a reduced precision arithmetic with the aim to decrease the necessary model size and computing time yet maintain the highest possible accuracy. The use-case is the reduction of the required hardware resources for future Large Hadron Collider (LHC) detector simulations at CERN.

The paper "Adaptive Sampling for Weighted Log-Rank Survival Trees Boosting" is the extension of a paper presented at ICPRAM 2022. The field of survival analysis tackles the problem of predicting the probability and time of the occurrence of an event. Survival decision tree models have strong interpretability and can evaluate the importance of predictors, but they demonstrate inferior performance in comparison to classical Cox proportional hazards models. These in turn suffer from the assumption of non-overlapping survival functions, which seldom hold on real data. The paper proposes a new boosting of the survival decision tree model that uses adaptive sampling and weighted log-rank split criteria.

The ICPRAM 2022 paper "Exploiting Temporal Coherence to Improve Person Re-identification" tackles the interesting problem of person re-identification in long-term scenarios. The use-case is whole-body runner re-identification during ultra-running competitions, where the task suffers from different illumination conditions, changes of clothing and/or accessories like backpacks, caps, and sunglasses. This paper explores integrating these cues with the spatio-temporal context information present in the competition live track system.

The paper "Perusal of Camera Trap Sequences Across Locations" (ICPRAM 2022) deals with the interesting problem of efficiently handling camera trap sequences in video analysis related to ecological conservation. In particular, the paper proposes a pipeline for wildlife detection and species recognition to expedite the processing of camera trap sequences. The proposed pipeline consists of the three stages of empty frame removal, wildlife detection, and species recognition and classification, which mostly rely on the use of the recently spreading visual transformers.

The paper "Gesture Recognition and Multi-modal Fusion on a New Hand Gesture Dataset" (ICPRAM 2022) presents a new hand gestures dataset and a baseline set of experiments using state-of-the-art sequence classifiers and the new dataset as a benchmark. The dataset consists of about 100,000 samples, grouped into six classes of typical and easy-to-learn hand gestures. The dataset was recorded using two independent sensors, allowing experiments on multi-modal data fusion at several depth levels.

The last paper from ICPRAM 2022 that closes this book is "Retinotopic Image Encoding by Samples of Counts". The generative approach proposed treats the image coding within generative models as a special case of the classical statistical problem of probability distribution density estimation, which is restricted to the class of parametric estimation procedures. In particular, the authors propose to use the model of a parametric mixture of simple distribution components.

While the theoretical scene seems presently dominated by deep models, and especially by generative ones, the introduced works also demonstrate the wide range of possible applications. We hope that researchers, engineers, and practitioners in the areas of pattern recognition and machine learning will find this book of interest.

Last but not least, we would like to thank all the authors for their contributions and also the reviewers, who have made this publication possible and helped ensure its quality.

February 2021

Maria De Marsico
Gabriella Sanniti di Baja
Ana Fred

Organization

Conference Chair

Ana Fred   Instituto de Telecomunicações and University of
                                        Lisbon, Portugal

Program Co-chairs

Maria De Marsico   Sapienza Università di Roma, Italy
Gabriella Sanniti di Baja   Institute of Cybernetics, Italian National Research
                                        Council (CNR), Italy

Program Committee

Served in 2021

Alessandra Lumini   Università di Bologna, Italy
Ana Sequeira   INESC TEC, Portugal
Andrea Abate   University of Salerno, Italy
Angelo Genovese   Università degli Studi di Milano, Italy
Antanas Verikas   Halmstad University, Sweden
Antonio Bandera   University of Málaga, Spain
Ashraf Abdel Raouf   Misr International University, Egypt
Bertrand Kerautret   Université Lumière Lyon 2 - LIRIS, France
Carlo Sansone   University of Naples Federico II, Italy
Dara Pir   Guttman Community College, City University of
                                        New York, USA

Fernando Rubio   Universidad Complutense de Madrid, Spain
Genny Tortora   Università degli Studi di Salerno, Italy
Gerald Schaefer   Loughborough University, UK
Haluk Eren   Firat University, Turkey
J. Salvador Sánchez   Universitat Jaume I, Spain
José Saavedra   Orand S.A, Chile
Juan Luo   George Mason University, USA
Lisimachos Kondi   University of Ionnina, Greece
Luciano Dutra   National Space Research Institute of Brazil -
                                        INPE, Brazil

Marco La Cascia   Università degli Studi di Palermo, Italy
Mario Köppen   Kyushu Institute of Technology, Japan

Mario Vento	Università degli Studi di Salerno, Italy
Markus Koskela	CSC - IT Center for Science Ltd., Finland
Miguel Coimbra	University of Porto, Portugal
Mulin Chen	Xidian University, China
Pascal Matsakis	University of Guelph, Canada
Petra Gomez-Krämer	La Rochelle University, France
Pornntiwa Pawara	University of Groningen, Netherlands
Rafel Rumi	University of Almería, Spain
Rebeca Marfil	University of Malaga, Spain
Reyer Zwiggelaar	Aberystwyth University, UK
Ricardo Torres	Norwegian University of Science and Technology, Norway
Rocio Gonzalez-Diaz	University of Seville, Spain
Shengkun Xie	Toronto Metroplitan University (Formerly Ryerson University), Canada
Simone Scardapane	Sapienza University of Rome, Italy
Stephen Pollard	HP Labs, UK
Yonggang Lu	Lanzhou University, China

Served in 2022

Josep Llados	Universitat Autònoma de Barcelona, Spain
Laurent Heutte	Université de Rouen, France

Served in 2021 and 2022

Adam Krzyzak	Concordia University, Canada
Akinori Ito	Tohoku University, Japan
Alfred Bruckstein	Technion, Israel
Andrea Bottino	Politecnico di Torino, Italy
Antonio-José Sánchez-Salmerón	Universitat Politecnica de Valencia, Spain
Asmir Vodencarevic	Novartis Pharma GmbH, Germany
Bernard Gosselin	University of Mons, Belgium
Bob Zhang	University of Macau, Macau
Boris Mirkin	National Research University Higher School of Economics, Russia
Daniel Riccio	University of Naples Federico II, Italy
Delia Mitrea	Technical University of Cluj-Napoca, Romania
Duane Edgington	Monterey Bay Aquarium Research Institute, USA
Edmondo Trentin	Università degli Studi di Siena, Italy
Eduardo Lleida	Universidad de Zaragoza, Spain
Elena Marchiori	Radboud University, Netherlands
Ernest Valveny	Universitat Autònoma de Barcelona, Spain

Eulalia Szmidt	Systems Research Institute, Polish Academy of Sciences, Poland
Friedhelm Schwenker	University of Ulm, Germany
George Azzopardi	University of Groningen, The Netherlands, and University of Malta, Malta
Ghada Zamzmi	National Institutes of Health, USA
Giorgio Fumera	University of Cagliari, Italy
Jakub Nalepa	Silesian University of Technology, Poland
Javier Calpe	Universitat de València, Spain
Javier Lorenzo-Navarro	Universidad de Las Palmas de Gran Canaria, Spain
Jean-Louis Dillenseger	Université de Rennes 1, France
Jing-Hao Xue	University College London, UK
José Alba	University of Vigo, Spain
Kalman Palagyi	University of Szeged, Hungary
Kidiyo Kpalma	Institut National des Sciences Appliquées de Rennes, France
Kouichi Hirata	Kyushu Institute of Technology, Japan
Laurent Wendling	Paris Cité University, France
Luiza de Macedo Mourelle	State University of Rio de Janeiro, Brazil
Malgorzata Kretowska	Bialystok University of Technology, Poland
Marco Maggini	University of Siena, Italy
Marco Muselli	Consiglio Nazionale delle Ricerche, Italy
Marek Kretowski	Bialystok University of Technology, Poland
Mark Nixon	University of Southampton, UK
Markus Goldstein	Ulm University of Applied Sciences, Germany
Mayer Aladjem	Ben-Gurion University of the Negev, Israel
Michal Haindl	Institute of Information Theory and Automation, Czech Republic
Michele Scarpiniti	Sapienza University of Rome, Italy
Mikael Nilsson	Lund University, Sweden
Mikhail Petrovskiy	Lomonosov Moscow State University, Russia
Mita Nasipuri	Jadavpur University, India
Modesto Castrillon-Santana	Universidad de Las Palmas de Gran Canaria, Spain
Monica Bianchini	University of Siena, Italy
Monique Thonnat	Inria, France
Mu-Chun Su	National Central University, Taiwan, Republic of China
Muhammad Marwan Muhammad Fuad	Coventry University, UK
Oriol Ramos Terrades	Universitat Autònoma de Barcelona, Spain
Paolo Addesso	Università degli Studi di Salerno, Italy

Paula Brito	Universidade do Porto, Portugal
Pavel Zemcik	Brno University of Technology, Czech Republic
Philippe Ravier	University of Orléans, France
Rahib Abiyev	Near East University, Turkey
Ramón Mollineda Cárdenas	Universitat Jaume I, Spain
Sarangapani Jagannathan	Missouri University of Science and Technology, USA
Sean Holden	University of Cambridge, UK
Sivaramakrishnan Rajaraman	National Library of Medicine, USA
Slawomir Wierzchon	Polish Academy of Sciences, Poland
Sotiris Kotsiantis	University of Patras, Greece
Stefano Berretti	University of Florence, Italy
Su-Yun Huang	Academia Sinica, Taiwan, Republic of China
Vijayakumar Bhagavatula	Carnegie Mellon University, USA
Vincenzo Piuri	Università degli Studi di Milano, Italy
Xiaoyi Jiang	University of Münster, Germany
Yago Diez	Yamagata University, Japan
Yaokai Feng	Kyushu University, Japan
Yoshito Otake	Nara Institute of Science and Technology, Japan
Young-Koo Lee	Kyung Hee University, South Korea
Yuji Iwahori	Chubu University, Japan

Additional Reviewers

Served in 2021

| Pietro Bongini | University of Siena, Italy |

Served in 2022

| Rita Delussu | University of Cagliari, Italy |

Invited Speakers

2021

Julian Fierrez	Universidad Autonoma de Madrid, Spain
Cornelia Fermüller	University of Maryland, USA
Marco Gori	University of Siena, Italy

2022

Mihaela van der Schaar	University of Cambridge, UK
Krystian Mikolajczyk	Imperial College London, UK
Tinne Tuytelaars	KU Leuven, Belgium
Nicu Sebe	University of Trento, Italy

Contents

Theory and Methods

Reduced Precision Research of a GAN Image Generation Use-case 3
  Florian Rehm, Vikram Saletore, Sofia Vallecorsa, Kerstin Borras,
  and Dirk Krücker

Similarity Constrained Conditional Generative Auto-encoder
with Generalized Dilated Networks 23
  Jan Niclas Reimann, Bhargav Bharat Shukla, Andreas Schwung,
  and Steven X. Ding

Retinotopic Image Encoding by Samples of Counts 52
  Viacheslav Antsiperov and Vladislav Kershner

Gesture Recognition and Multi-modal Fusion on a New Hand Gesture
Dataset 76
  Monika Schak and Alexander Gepperth

Adaptive Sampling for Weighted Log-Rank Survival Trees Boosting 98
  Iulii Vasilev, Mikhail Petrovskiy, and Igor Mashechkin

Applications

Forecasting Overtime Budgets for Naval Fleet Maintenance Facilities
Using Time-Series Analysis During Transient System States 119
  Charith Gunasekara, Lise Arseneau, and Cheryl Eisler

Exploiting Temporal Coherence to Improve Person Re-identification 134
  Oliverio J. Santana, Javier Lorenzo-Navarro, David Freire-Obregón,
  Daniel Hernández-Sosa, José Isern-González,
  and Modesto Castrillón-Santana

Perusal of Camera Trap Sequences Across Locations 152
  Anoushka Banerjee, Dileep Aroor Dinesh, and Arnav Bhavsar

Author Index 175

Theory and Methods

Reduced Precision Research of a GAN Image Generation Use-case

Florian Rehm[1,2]([✉])(iD), Vikram Saletore[3](iD), Sofia Vallecorsa[1](iD), Kerstin Borras[2,4](iD), and Dirk Krücker[4](iD)

[1] CERN, Esplanade des Particules 1, Geneva, Switzerland
florian.matthias.rehm@cern.ch
[2] RWTH Aachen University, Templergraben 55, Aachen, Germany
[3] Intel, 2200 Mission College Blvd., Santa Clara, CA, USA
[4] DESY, Notkestraße 85, Hamburg, Germany

Abstract. In this research a deep convolutional Generative Adversarial Network (GAN) model is post-training quantized to a reduced precision arithmetic for a complex High Energy Physics (HEP) use-case. This research is motivated by the aim to decrease the necessary model size and computing time for reducing the required hardware resources for future Large Hadron Collider (LHC) detector simulations at CERN. However, in order to interpret the measured physics results, the detector simulations have to maintain the highest possible accuracy. Therefore, the quantized model is not only in detail analyzed in terms of hardware resource consumption but additionally comprehensively evaluated in terms of the achieved physics accuracy. We report that we achieve with the quantized model a 3.0x speed-up versus the initial model on modern CPUs. Furthermore, we investigate several new physics accuracy metrics to demonstrate that the accuracy does not significantly decrease due to the quantization process. Reduced precision computing for classification problems is already adequately studied, however, this is not the case for more complex image generation problems as we require for our use-case of detector simulations in this research. By using the Intel Neural Compressor, the quantization is performed in an iterative process. Neural Compressor automatically quantizes only the parameters of the neural network which do not decrease the accuracy of the model regarding a predefined accuracy metric. In our research we post-training quantize the GAN model from the 32-bit format down to 8-bit format.

Keywords: Reduced precision computing · Generative Adversarial Networks (GAN) · Neural Compressor · HEP simulation · Quantization

1 Introduction

Deep Learning (DL) achieves today in many applications almost human efficiency and replaces traditional computing algorithms because of its higher effectiveness [10]. Due to increasingly deeper neural networks for achieving higher accuracies the simulation time rises. In some application fields, for example on mobile edge devices, computational power is limited. However, for increasing model sizes hardware resource reduction additionally becomes relevant for applications on CPU and GPU architectures. This

© Springer Nature Switzerland AG 2023
M. De Marsico et al. (Eds.): ICPRAM 2021/2022, LNCS 13822, pp. 3–22, 2023.
https://doi.org/10.1007/978-3-031-24538-1_1

is also the case for High Energy Physics (HEP), where enormous amounts of data have to be simulated within short time frames. A simple approach to reduce hardware needs for a trained DL model represents the quantization of the neural networks into a reduced precision.

In this section we provide an introduction to reduced precision computing, to the physics use-case of detector simulations and to the generative model used for the simulation. In the subsequent chapter we introduce the quantization tool Neural Compressor applied in our study. In the following we evaluate the quantized model and compare the results to the initial not quantized model and to the classical Monte Carlo simulation we aim to replace. Finally, at the end we draw conclusions.

1.1   Reduced Precision Computing

In reduced precision computing higher throughput during inference and reduced memory storage are achieved by shrinking the activations and weights of the trained neural network to a lower precision. The process of converting numbers from higher to lower formats is named quantization. The standard number format in machine learning is floating point 32 (float32) or also named single precision [11]. It uses 4 bytes or 32 bits for representing a single number. The float32 format utilizes one bit for the sign, eight bits for the exponent and 23 bits for the fraction. For illustration, the largest possible number of float32 is $2^{128} = 3.40 \cdot 10^{38}$. The format in which we quantize our model is integer-8 (int8) which uses, as the name indicates, 8 bits (or 1 byte) for representing a single number. This is a quarter of the number of bits of the float32 format. Integer means only whole numbers without fractions are allowed. The number can be either signed int8 (sint8) with a range of $[-128, 127]$ or unsigned int8 (uint8) with a range of $[0, 255]$. Figure 1 shows graphically the float32, sint8 and uint8 format for comparison.

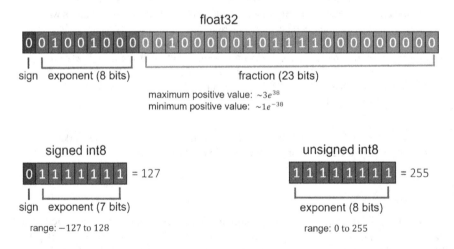

Fig. 1. (top) Representation of the float32 format, (bottom left) signed int8 format and (bottom right) unsigned int8 format. The int8 formats require only a quarter of the bits of the float32 format.

Related Work Quantization: There exist two different quantization techniques. The first is post-training quantization, where the model is trained in full precision and quantized after training. For quantizing into a lower precision, the use of a calibration dataset is necessary to calculate the maxima of the weights and activations. Because post-training quantization represents a straightforward approach which leads in most cases to a satisfying level of accuracy, it is commonly the first approach chosen [30]. Hence, we employ this approach for the study in this paper and explain post-training quantization in the following section in more detail. Typically, models are trained in float32 precision on multi-node clusters, which are then quantized to lower precision formats and loaded on light-weight devices, such as edge-devices. Ref. [13] describes the quantization process for quantizing convolutional networks down to as low as 4-bits with only using small calibration data sets while avoiding overfitting. Another research was carried out in Ref. [5], where they applied different quantization techniques for quantizing pre-trained state-of-the art neural networks down to 4-bit while only losing a few percent of accuracy.

An alternative method to the one explained above, is named quantization-aware training. It directly trains models using lower precision formats. However, quantization-aware training is more complicated and, therefore, not used in this study. The benefit of quantization-aware training is that the quantization errors which decrease the total loss of the model can be reduced by optimizing the parameters accordingly. In Ref. [10], they compare the training of neural networks in various precision formats. In Ref. [29], they demonstrate a $2 - 4x$ training time improvement with the use of 8-bit floating point numbers above today's 32-bit systems.

A further step is to train models with mixed precision formats using both, float32 and int8 for example. In this case, some weights are represented in int8 format to achieve a speed-up, but others are kept in float32 to maintain the level of accuracy. Choosing which weights are represented in higher or lower precision is performed by intelligent algorithms. Various studies have evaluated mixed precision for training [16,17]. In particular Ref. [20] employs successfully a mixed precision approach for the training of GANs.

In most cases, quantization techniques are targeted to int8 format. Because int8 is supported by most modern hardware and maintains almost the equivalent level of accuracy, in comparison to even lower precision formats which cannot maintain the accuracy [14]. Most of the existing benchmarks represent classification problems [31], while in our case, we require an accurate description of generated image data.

The content of this paper represents an extension to the previous publication from Ref. [25] where we published some preliminary results. The results presented in this paper demonstrate the conclusive outcomes of our quantization research which are in more detail explained and investigated here. The good physics accuracy of our quantized model is confirmed with new optimized physics validation metrics which offer a deeper insight and understanding of the model. For performing the quantization, we use the quantization tool Neural Compressor as previously. Paying tribute to the hardware evolution, we run the inference benchmark tests on the latest Intel Ice Lake hardware just published a few months ago and accomplish an almost double as large speed-up than previously seen. Therefore, most of the content provided in this paper is novel compared to the previous research in Ref. [25].

Mathematics Behind Quantization: In the post-training quantization process we want to scale down the number from a range of $[-3.4 \cdot 10^{38}, 3.4 \cdot 10^{38}]$ to a range of $[0, 255]$ for using uint8 format or $[-127, 128]$ for using sint8. In order to do this, first the maximum absolute numbers of the weights B_w and activations B_α of the neural network are identified. These are required to obtain the boundaries for calculating scaling factors which allow to use the full range of the int8 numbers after quantization. To measure the maxima B_w and B_α, it is required to run inference with a few batches of the training data with the trained model before quantizing it. The scaling factor can then be calculated with

$$S_{unsigned,(w,\alpha)} = \frac{255}{B_{(w,\alpha)}} \tag{1}$$

for using the full range of uint8 numbers ($\in [0, 255]$) or

$$S_{signed,(w,\alpha)} = \frac{127}{B_{(w,\alpha)}} \tag{2}$$

for sint8 numbers ($\in [-127, 128]$). The calculation is done independently for the weights w and the activations α. The quantized weight I_i can then be calculated with

$$Q_{wi} = round(S_w \cdot I_i), \tag{3}$$

where the results are rounded to the closest integer number. The index i denotes the ith weight. Similarly, as the weights the quantized activations can be calculated.

1.2   High Energy Physics Simulations

The Large Hadron Collider (LHC) at CERN in Switzerland is the largest particle accelerator in the world ever built for accelerating charged particles to the maximum possible amount of energy. The high energetic particles are then collided with a second counter-rotating beam in huge particle physics detectors to study the particle's properties and new physics after the interaction. These detectors are constructed by multiple sub-detectors, each serving different purposes in order to measure all the quantities of the particles created in the collision, namely energy, momentum, angle and charge. These quantities are necessary to determine the type of the measured particles as well as to reconstruct the interaction. This is a complex process, and its analysis is borne by simulations deriving the accuracy parameters of the reconstruction. Nowadays, the Geant4 toolkit [2] is employed for detailed simulation of the interactions while the particles pass through the matter of the detectors. Geant4 is a complex tool consisting of detailed Monte Carlo simulations for reproducing the known physics processes.

HEP Monte Carlo simulations are computing resources demanding tasks because of the complex underlying physics processes. In particular, calorimeters, one type of sub-detectors, are built from a dense material in a highly granular geometry with many sensors which leads to elaborated and extended simulations. At the world leading HEP institution CERN detector simulations require currently more than half of the world-wide (LHC) grid resources [3]. In the future LHC high-luminosity phase, presently scheduled for 2027, the required amount of simulated data is expected to be around hundred times larger than today due to more simultaneous particle collisions (called pile-up) and calorimeter detectors with a novel grade of high granularity. This exceeds drastically the expected computational resources even with taking the technology improvement into account [4]. For mastering this obstacle, intense research is ongoing for developing alternative approaches to the recent Monte Carlo simulation. The major difficulty for alternative simulation approaches lies in the required high level of accuracy in order to successfully interpret the measured physics results with only small uncertainties.

In the recent years deep learning simulations were increasingly adopted in HEP because they tend to work even better when more data is available. Once deep learning models are trained in a complex and tedious process, their application or inference is fast and scaling-up well. As a first step and serving as a proof of concept, in this research the most hardware consuming part of the detector simulations, the calorimeter simulations, are replaced by deep learning methods. For this several prototypes based on deep generative models are tested [19, 21, 24].

Electromagnetic Calorimeters: Calorimeters are devices which are operated in HEP detectors to measure particle energies. In this study we focus on electromagnetic calorimeters which are a sub-type of calorimeters measuring electrons and photons as primary particles. These primary particles enter the massive detector material and produce a cascade of secondary particles while they pass through the detector. The secondary particles create then other particles with the same mechanism. This leads to a particle shower inside the calorimeter. The dominant underlying physics processes which cause secondary particles in matter are bremsstrahlung for electrons and pair production for photons. When the energy of the particles falls below a certain threshold, the energy of the secondary particles is absorbed and measured by the calorimeter sensors.

The calorimeter in this study represents an illustrative prototype of a future high-granularity calorimeter. A higher granularity corresponds to more detection and measurement volumes and therefore to a higher detector precision. The calorimeter is built as a three-dimensional volume and measures the absorbed energy with $25 \times 25 \times 25$ cells summing up in a total of 15625 pixels. A demonstrative example shower image is shown in Fig. 2. The primary particles enter the calorimeter in the middle of the x- and y- direction and the particle shower is evolving in the z-direction named the calorimeter depth.

It is worth to mention some particularities of the calorimeter training data. The shower images are very sparse, only a minor fraction of pixels receives some energy deposition. Furthermore, the energy range of the pixel entries can vary in more than six orders of magnitude. In this study $200\,000$ shower images are used (90% training data, 10% test data) with primary particle energies in the range of $E_p = [100, 500]$ GeV. As we will explain in Sect. 3, for validation a second data set is used consisting of $40\,000$ additional images. The data of both data sets is generated by Geant4 with detailed Monte Carlo simulation and is public available in Ref. [22].

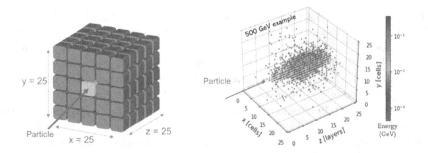

Fig. 2. (left) Schematics of the 3D calorimeter volume and (right) an example shower image for a primary particle with 500 GeV energy.

1.3 Deep Convolutional Generative Adversarial Networks

The simulation approach employed in this paper are Generative Adversarial Networks (GANs). GANs were first introduced by Ian Goodfellow [9] in 2014 and represent today a state-of-the-art technique for image generation. The GAN principle is that two models or two neural networks are carrying out an adversarial role based on game theory. The generator network G tries to fool the discriminator network D by sending fake images which are labeled as true images (training images). G generates the fake images x from a random latent variable z. The discriminator on the other hand, tries to distinguish between real data (images from the training data set) and fake data x (generated images). The training goes in alternating turns. In the first turn the generator tries to improve its strategy to make its generated images closer to the one from the training data. In the subsequent turn the discriminator improves its network to correctly distinguish between the generated and the training data before the generator is trained again, etc. The training is successful, when the discriminator is no more able to distinguish between the original images and synthetic results producing a classification prediction y of 50% for each class. The two GAN models, generator and discriminator, are typically parameterized by deep neural networks which are trained simultaneously by the following objective function:

$$\min_{G} \max_{D} V(G, D) = \mathbb{E}_{x \sim p_{data}(x)}[\log D(x)] + \mathbb{E}_{z \sim p_z(z)}[\log(1 - D(G(z)))]. \quad (4)$$

Phased differently, the generator and discriminator play a two-player min-max game, the generator tries to minimize the objective function in order to produce the generated images more similar to the training data until they are indistinguishable. The discriminator, on the other hand, maximizes the objective function to distinguish between generated data and training data. The GAN principle is demonstrated in Fig. 3.

Conv2D Neural Network Architectures: As we apply the GAN to an image generation task, the neural networks primarily consist of convolutional layers. Therefore, the used model represents a Deep Convolutional GAN (DC-GAN). In the following the neural network architectures of the generator and discriminator network are briefly introduced because the contained layers are important for the later quantization process.

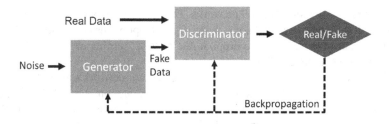

Fig. 3. The GAN principle with the two neural networks, the generator and the discriminator, which are trained in an adversarial strategy.

The generator architecture is shown in Fig. 4. The output of the generator is a 3D image with the dimension $25 \times 25 \times 25$ pixels, the same dimension as the training images. The input into the generator represents a latent vector consisting of 200 random numbers drawn from a uniform distribution. It is multiplied by the input energy E_p of the primary particle which represents the only dependency of the generated showers. After placing a dense layer, the input tensor is reshaped to a 3D volume with the dimensions $5 \times 5 \times 5$. Usually, one would use 3D convolutional (Conv3D) layers for processing 3D images. However, as Conv3D layers are yet unsupported in the quantization tool, we developed a smart neural network architecture consisting of 2D convolutional (Conv2D) layers. As shown in a previous publication in Ref. [24], with this approach we were able to generate successfully 3D images with only Conv2D layers. Furthermore, in terms of physics accuracy as well as computation time, the new model clearly outperformed the previous Conv3D architecture.

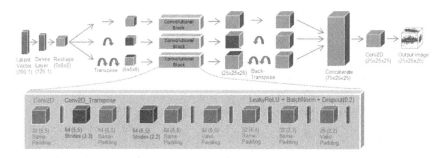

Fig. 4. The Conv2D generator architecture [25].

In addition to the Conv2D layers, Conv2D_Transpose layers are responsible for upsampling the volume to a larger size to produce the required dimensions at the output. Within the convolutional block after each Conv2D layer a LeakyReLU activation function is applied in order to introduce none-linearities into the network to allow to learn complex patterns from the data. The LeakyReLU is followed by a BatchNorm layer which re-centralizes and re-scales the data to execute the parameters of the previous convolutional layer more stable. BatchNorm layers possess a high importance, especially in helping to converge during the training process and to stabilize the training of deep neural networks. The last layers comprise Dropout layers, which represent a method of regularization of neural networks. They reduce the possibility of overfitting by turning off, named "dropout", some nodes from the previous convolutional layer by setting its output to zero. We use a dropout rate of 20%, turning off randomly 20% of the nodes to prevent mode collapse.

Initially, we wanted to implement within the generator network standard ReLU activation functions instead of LeakyReLU ones. However, our complex GAN model combined with the sparse shower images as training data do not allow to use the plain ReLU activation functions. Training our model with ReLU functions leads to convergence to zero gradients everywhere after a few optimization steps. This issue is known as the dying ReLU problem and appears in deep network architectures where many (or all) neurons output the values 0. When all the gradients of the network return zero it is unable to learn further in the backpropagation process and the full model remains unrecoverable [15]. When using LeakyReLU activation functions this problem does not appear and we are able to train the model and to generate meaningful images. Other approaches which typically combat the dying ReLU problem are to decrease the learning rate and/or to adapt the weight initialization. However, these techniques do not resolve the dying ReLU problem for our model, most probably because of the sparse shower images.

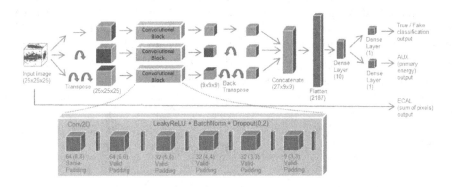

Fig. 5. The Conv2D discriminator architecture [25].

The corresponding discriminator architecture similarly operates with Conv2D layers and is shown in Fig. 5. The inputs of the discriminator are either generated images (called fake images) or images from the training dataset (called true images). The discriminator outputs three values which represent the loss values for the optimization of the neural network parameters during training:

- The first loss value is the typical GAN true/fake probability L_{TF} [9]. It predicts how likely it is, that the input image represents a true image (from the training set) or contrary, how likely it is, that the input image represents a fake image. The true/fake loss is calculated using the binary cross entropy [18] after applying a sigmoid activation function.
- The second loss, named AUX (for AUXiliary loss) L_{AUX}, represents the result of a regression task used to model the relation between the initial particle energy E_p

and the discriminator estimate from the images using a separate dense layer. It is implemented as a Mean Absolute Percentage Error (MAPE) [26].

- The third discriminator output comes from a Lambda layer, a custom layer calculating the sum over the pixels of the input image and corresponds to the total energy of the input image. It is entitled ECAL L_{ECAL} and evaluates if the generated image contain the correct amount of energy as expected from the training dataset. The loss is calculated by the MAPE likewise. The ECAL loss does not depend on the discriminator network and therefore only updates during training the parameters of the generator network and not the ones of the discriminator network.

The total loss L_{total} of the discriminator is calculated by multiplying each of the three losses with a scalar weight w as hyperparameter and summing them up:

$$L_{total} = w_{TF} \cdot L_{TF} + w_{AUX} \cdot L_{AUX} + w_{ECAL} \cdot L_{ECAL} \qquad (5)$$

In addition to the true/fake loss, the two losses AUX and ECAL are required to help the GAN to converge faster with fewer oscillations during training and to reach higher accuracy's. The loss weights are chosen to equally weigh each of the loss terms:

- $w_{TF} = 6.0$
- $w_{AUX} = 0.2$
- $w_{ECAL} = 0.1$

The aim of the trained generator is to provide the corresponding particle detector output for a specific particle type and energy. The discriminator is only required for training the generator and, therefore, for the inference process not further needed. Hence, in the following reduced precision research the focus lies exclusively on quantizing the generator network.

2   Quantization Tool

Because reduced precision computing is nowadays an established technique of decreasing hardware requirements there exist already many quantization tools. The primary constraint for quantizing the above explained generator network into lower precision is that in most of the tools, not all layers which are present in our neural network, are integrated. In Sect. 1.3, we already indicated that Conv3D layers are not supported in any quantization tool because they are in general not very frequently applied. In order to explore and study reduced precision computing we had to come up with a new neural network architecture which uses the more common Conv2D layer as described in Sect. 1.3. However, additionally there are other layers in our network which cause difficulties in finding the appropriate quantization tool. These comprise Conv2D_Transpose layers and LeakyReLU activation functions which are at this time not supported in most quantization frameworks. The reason for this is that quantization in deep learning is typically applied to classification problems which employ simpler neural network outputs (probability values of classes) than compared to the complex 3D shower images from our network. These classification tasks use the standard ReLU activation functions instead of LeakyReLU's. However, as explained in Sect. 1.3, we need the LeakyReLU activation functions to prevent the dying ReLU problem. Additionally, ordinary classification problems have no need for Conv2D_Transpose layers, because there is no upsampling in image size required. Therefore, also Conv2D_Transpose layers are in most quantization tools at this time not implemented.

For using quantized LeakyReLU layers an additional problem appears in TensorFlow (TF). TF supports by default only uint8 (unsigned int8) operations and not sint8 (signed int8) operations. The sint8 operations are not needed for classification networks with standard ReLU activations and, therefore, sint8 are by default not yet implemented in TF. But, the latter ones are necessary for our model since LeakyReLU functions can generate negative values. Therefore, because we need LeakyReLU layers in the network, there is no other way than implementing sint8 operations within a patched TF version, as we will discuss in the following.

Intel Neural Compressor: The Intel Neural Compressor [1] (formerly known as Intel Low Precision Tool, LPOT) is an open-source python library for quantizing deep learning models and running low precision inference across multiple frameworks. It uses the Intel oneAPI Deep Neural Network Library (oneDNNL) [12], which contains building blocks for deep learning applications to improve the performance on Intel processors. In the previous research in Ref. [25], we were the first who applied the brand-new Neural Compressor tool on a real use-case and achieved good results. An overview of the Intel Neural Compressor infrastructure can be seen in Fig. 6.

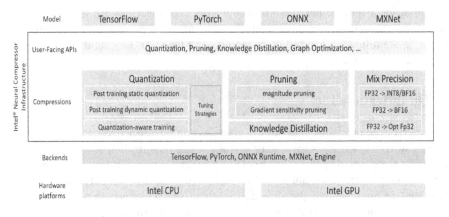

Fig. 6. Overview of the Intel Neural Compressor framework [1].

Neural Compressor is a post-processing quantization tool which automatically optimizes deep learning models with low-precision recipes during quantization in order to achieve the optimal objectives. The objectives for which Neural Compressor tunes are inference time, memory usage and model accuracy [1]. For achieving this, it searches for the best order of magnitude for the scaling factors for an entire layer instead of single pixels. A key feature is that Neural Compressor can drop single "outlier" bits which are far away from the other values within the corresponding layer and which would distort the results. With a recurrent refinement of the topology and dropping outliers, the

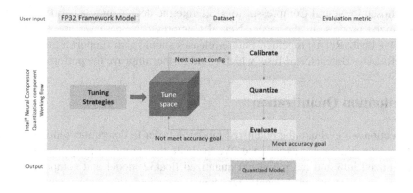

Fig. 7. The workflow of Intel Neural Compressor [1].

Neural Compressor quantization remains a challenging process instead of a one-step task compared to most other quantization tools. The quantization workflow of Neural Compressor that aids in increasing performance is displayed in Fig. 7. Furthermore, in the automatic quantization process Neural Compressor keeps a few nodes of the neural network in float32 precision in order to increase or to keep a high accuracy. Therefore, Neural Compressor is a post-processing mixed precision tool. This is another advantage compared to other simpler quantization tools because it can reach much higher accuracies due to the mixed precision. A comparison of Neural Compressor versus TensorFlow Lite was demonstrated in Ref. [25]. There the Neural Compressor int8 model performed in terms of accuracy much better than the TensorFlow Lite int8 model. For measuring the accuracy during the quantization process we use for our model a physics validation metrics which will be introduced in the evaluation Sect. 3.

As mentioned earlier, sint8 operations are by default not implemented in TF which we employ for our research. This leads to the fact that the recent quantization tools have not implemented quantized LeakyReLU functions. Therefore, in this study we use a customized TF version supporting sint8 operations which was provided by Intel. Furthermore, the Intel Neural Compressor team implemented for us the quantized LeakyReLU function into the Neural Compressor quantization tool. Neural Compressor with the patched TF version is at this time the only set-up which provides all required functionalities to quantize our model. The Neural Compressor team applied their quantization tool to popular neural network architectures, such as ResNet and Inception and achieved on average a two to three times speed-up in inference while only loosing less than a half percent in accuracy, see Ref. [8]. Due to the optimization process of Neural Compressor, they were even capable in some cases to increase the accuracy for the quantized model.

In many neural network architectures a set of layers is multiple times repeated. This is also the case in our generator network, where we have a Conv2D layer, followed by a LeakyReLU and a BatchNorm layer multiple time. By combining these often-repeated layers of a network one can significantly decrease the number of layers within the network and further speed-up the inference process. The layer fusion strategy can be chosen in Neural Compressor to accelerate the deep neural networks inference for quantized models. In the recent Neural Compressor release we are able to fuse [Conv2D + LeakyReLU] together. Future releases should also support a full [Conv2D + LeakyReLU + BatchNorm] fuse, which could further improve the performance.

3 Evaluation Quantization

In this section we evaluate the quantized model in order to determine which improvements or deterioration we experience. We evaluate the inference process of both, the new quantized int8 and the initial not quantized float32 model and compare them to the detailed simulations with Geant4, which are very computing resource intensive and need to be replaced. Inference represents the process of applying a neural network after training to the specific use-case for which it is designed. For the GAN model under study inference is the process when we generate a shower image with the generator model by providing an input energy E_p. It is very important, that the inference process occupies as less computational resources as possible while reaching the best achievable physics accuracy for interpreting the physics results of the generated showers. When a potential new model replaces the presently used Geant4 Monte Carlo simulations, it must produce shower images which are as close as possible to the Monte Carlo simulation in order to interpret the results physics-wise correctly. Therefore, we will go into the very details and discuss many different metrics which enable a better understanding of the physics performance. In the following we investigate first the computational performance and afterwards the physics accuracy.

3.1 Computational Evaluation

To evaluate the computational performance, we measure the inference time as well as the model memory occupancy of the int8 model and compare it to the float32 model. We aim for a significant improvement in order to make the calorimeter simulations as computing resource efficient as possible in order to scale-up.

The inference time depends on the model and on the hardware on which it runs. The hardware tests in this paper are performed on the brand-new Intel Ice Lake 2S Xeon 8380 processors which provide 40 cores and 80 threads. The Intel Ice Lake architecture supports reduced precision data formats and therefore we employ it to benchmark the quantized int8 model versus the float32 model. All tests were run including 100 warm-up batches.

We measure the number of generated showers per second for both models. For all hardware tests we use Python version 3.6.8, the customized TensorFlow version 2.3.0 from [25] and a batch size of 128. The inference is run with multiple configurations of cores and streams and the results are summarized in Fig. 8. We can see that for all configurations the int8 model generates more showers per second compared to the float32 model resulting in a higher throughput and a lower inference time. The goal is to reach the highest possible throughput. The number of showers per second rises for configurations with more cores and streams as expected (from left to right in Fig. 8). However,

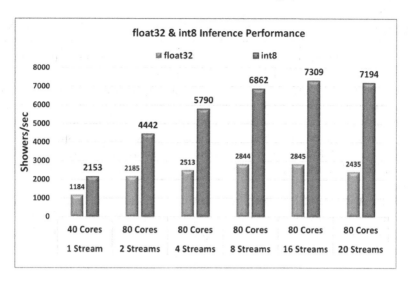

Fig. 8. The number of generated showers per second of the float32 and int8 model for multiple cores and streams configurations.

at the last configuration with 80 cores and 20 streams the number of showers per seconds drops slightly, most probably because the memory bandwidth limit is exceeded and the CPU becomes oversubscribed. This phenomenon was already experienced and discussed in the research in Ref. [25].

The best result in terms of number of showers per second is accomplished with the configuration of 80 cores and 16 streams where we are capable to generate 7 309 showers per second for the int8 model. Compared to the float32 model with 2 845 showers per second the int8 model is 2.6x faster in inference as summarized in Table 1. With this configuration the int8 model is 124 261 times faster than the Geant4 simulation which we aim to replace. For the Geant4 simulation time we refer to a previous measurement performed in Ref. [28], where it took 17 s for generating a single shower image on an Intel Xeon processor. The highest speed-up of int8 vs. float32 we achieve with the configuration of 80 cores and 20 streams where we measure a speed-up of 3.0x.

Table 1. The speed-up for the best configurations of the int8 model compared to float32 (denoted as int8/float32 scaling) and the speed-up of the int8 model versus Geant4 simulation.

Configuration	int8/float32 scaling	int8 speed-up vs. Geant4
80 Cores, 16 Streams	2.6x	124 261x
80 Cores, 20 Streams	3.0x	122 291x

The speed-up is a result of savings in memory bandwidth and faster computing with the int8 arithmetic. Because the number format of int8 is 4 times smaller than of float32 what leads to the expectation of a 4x speed-up. However, because through the quantization some more operations are added to the model and the computational time slightly increases. An example for an additional quantization operation at the beginning is the formatting of the input of the model from float32 down to int8. Additionally, the opposite appears at the end of the model, where a de-quantization layer de-quantizes the model output from int8 back to float32 format such that it can be processed by following algorithms. In comparison, the TensorFlow Lite team measures for different models inference speed-ups of around 3x [27] and the PyTorch team speed-ups in the order of 2 − 4x [23]. In the related work section of quantization we introduced already another study which achieved also speed-ups of 2 − 4x. These values agree with our measurement.

With the above results we set a new benchmark in terms of inference time for calorimeter simulations. In our previous research in Ref. [25] we reached only a 1.7x speed-up of int8 vs. float32 with a 67 000x speed up vs. Geant4. The new results in this paper are almost twice better than in the previous research.

Table 2. The hardware memory consumption of the int8 and float32 model.

Model	Memory Consumption
float32	8.083 MB
int8	3.571 MB

Next, we investigate the amount of hardware memory the model requires for both precision formats. As displayed in Table 2, the initial not quantized float32 model occupies 8.083 MB of hardware memory. Due to the quantization we were able to reduce the int8 model size to 3.571 MB. This results in an improvement of 2.26x in memory consumption for the int8 model with respect to the initial float32.

3.2   Physics Accuracy Evaluation

Performance evaluation of generative models remains a challenging task for which, depending on the specific applications, no uniform but several methods have been proposed [6]. Because calorimeter simulations are an image generation task, the majority of our validation metrics are visual. For evaluating the physics accuracy, we measure the energy patterns across the calorimeter volume focusing on specific quantities typically used to characterize calorimeter shower properties. For this study, we apply new physics validation methods to compare the generated shower images of the generator models to the ones of Geant4 in many possible aspects.

Example 3D Shower Images: As an initial investigation we compare the 3D shower images visually. For this a shower for the float32 and for the int8 model are generated for the same input vector. The example in Fig. 9 displays on the left the shower generated from the float32 model and on the right from the int8 model for a particle with a primary

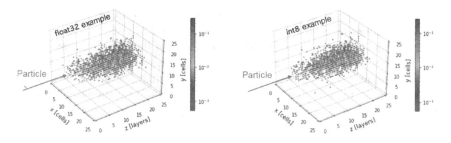

Fig. 9. 3D shower image for an $E_p = 400\,\text{GeV}$ example (left) for the float32 model and (right) for the int8 model.

energy of $E_p = 400\,\text{GeV}$. As expected, there is not much difference between the 3D shower images visible. Only looking into details at single pixels some minor differences become visible. In the 3D shower plots an energy cut for energies below $0.004\,\text{GeV}$ is applied in order to not overload the images and to make the core of the distribution visible.

Shower Shapes: With the shower shape metrics one can visually investigate the 2D projections of the energy shapes along the y- and z- axis averaged over $40\,000$ samples over the full energy range. The x-axis projections are similar to the ones of the y-axis due to the calorimeter geometry and therefore not shown. The validation data samples are, as the training data, generated by Geant4 and represent an independent data set used for all following validation metrics. The shower shape plots can be seen in Fig. 10 on a linear (left) and a logarithmic energy scale (right). The particle enters the detector orthogonally to its surface at the coordinates $x = 13$, $y = 13$ and $z = 0$. Therefore, in the transverse (x, y) plane larger energy depositions clusters are created around the middle of the image. The energy deposition along the z-axis rises until a peak because of the particle shower effect and then decreases because the secondary particle energies fall below the sampling threshold and become absorbed. One can see, that both, the float32 and the int8 model, are in the linear and logarithmic plot close to Geant4 except for a few pixels where the int8 model is slightly off. Therefore, the accuracy of the shower shapes of the int8 model is only marginally lower than for Geant4. However, it is astonishing that the quantized model works even for the extremely low energies at the distribution tails quite well, although it uses merely 256 numbers to represent the full pixel energy range. This high accuracy in the shower shape plots of the int8 model is possible due to the optimization process within Neural Compressor during quantization. In Ref. [25] we provided a comparison to models quantized by TensorFlow Lite without any optimization process. The TensorFlow Lite model performed especially at the low energy distribution tails bad.

MSE Single Validation Number: The shower shapes are the most important metrics to evaluate the generated shower images. Therefore, we created a validation number which provides the accuracy contained in the shower shape plots within one single number.

This number is calculated by the Mean Squared Error (MSE) from the 2-dimensional shower distributions pixel wise along the x-, y- and z-axis between the corresponding GAN model and Geant4. Afterwards, the MSE results of the three axes are averaged to obtain a single number. This validation number possesses a high significance and importance, because it is employed during the GAN training to evaluate the generator model accuracy for the single epochs. Additionally, the MSE validation number is used in Neural Compressor as accuracy metrics for which the model is optimized during quantization.

Table 3. The mean MSE and standard deviation (STD) of ten inference runs.

Model	Mean of MSE	STD of MSE
float32	0.0620	0.0012
int8	0.0529	0.0018

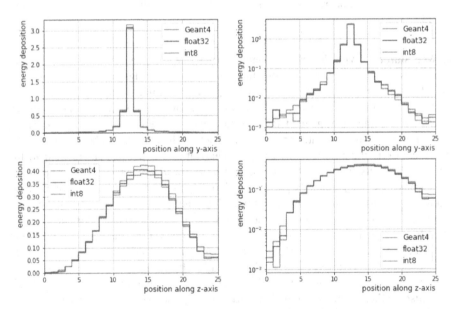

Fig. 10. The generated shower shape plots (left) on a linear and (right) on a logarithmic energy scale. On the horizontal axis the averages over the single pixels along the corresponding axis (top y-axis, bottom z-axis) are shown and on the vertical axis the contained energy.

The MSE values of the models are provided in Table 3. The table shows the mean and the standard deviation (STD) of the MSE for running ten times inference with 40 000 samples each. The MSE for the quantized int8 model (0.0529) is lower than for the float32 model (0.0620), this results in a higher accuracy because the MSE represents an error and, the lower the value the better the accuracy. Usually, the accuracy of models drops after quantization. In our case it was possible to accomplish a even higher accuracy because we optimized for the MSE single validation number in the quantiza-tion process within Neural Compressor. As already mentioned above, this optimization is key for maintaining a reasonable accuracy even after quantizing the model.

Mean Energy Deposition: The mean energy deposition inside the calorimeter is shown in Fig. 11 (left) with respect to the incident electron energy E_p for the full energy range. This metrics was proposed in Ref. [7]. To minimize statistical effects, the input energies for the GAN models are the same as for the Geant4 events in the validation data. An almost linear dependency is seen between the incident electron energy E_p and the measured mean energy μ_{90}, for which only the 90% core of the distribution is taken into account in order to discard boundary effects. In the upper plot the mean energy μ_{90} and in the lower plot the relative error is displayed. It can be seen, that the int8 model performs slightly worse than the float32 model because of the higher relative error. At 100 GeV the discrepancy of the int8 model is largest with a deviation of 2% versus Geant4. However, it should be noted, that deviations up to a few percent compared to Geant4 are expected for fast simulations as the GAN approaches. Therefore, the int8 model performance is acceptable.

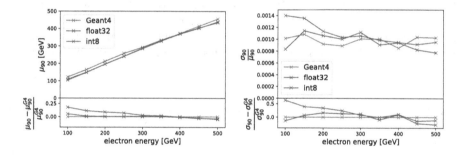

Fig. 11. Deposited energy mean μ_{90} (left) and relative width σ_{90}/μ_{90} (right). The lower panels show the relative deviation to Geant4.

In the right plot in Fig. 11 the relative width σ_{90}/μ_{90} is shown, where σ_{90} is the standard deviation of the 90% core of the distribution. For lower energies the int8 model performs worse than the float32 model with deviations up to 5% between the int8 model and Geant4.

Energy Sum: The energy sum metrics represents the sum of the energy of all pix-els of the shower image and corresponds to the total measured energy for one event. In Fig. 12 the energy sum is displayed for the discrete input energy events $E_p = [100, 300, 500]$ GeV. On the x-axis is the measured energy in MeV shown and on the y-axis the number of counts within an energy bin in an arbitrary unit.

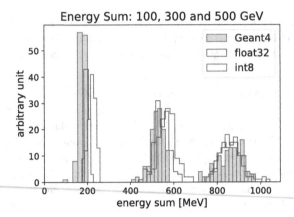

Fig. 12. Total deposited energy. A 10^{-6} GeV single cell energy threshold is applied for taking the detector resolution into account. The left peak corresponds to 100 GeV, the middle to 300 GeV and the right to 500 GeV.

The distributions of the float32 and int8 model are close to the ones from Geant4. However, the 100 GeV and 300 GeV event means are shifted slightly to higher energies. The distribution parameters mean μ and standard deviation σ are shown in Table 4. One can see there is a small shift in the energy sum distribution due to the quantization. However, already the initial float32 model shows some shift in the distributions with respect to Geant4.

Table 4. The means μ and the standard deviations of the energy distributions are shown for each model.

Model	$\mu_{100\,GeV}$	$\mu_{300\,GeV}$	$\mu_{500\,GeV}$
Geant4	175 ± 14	523 ± 32	856 ± 57
float32	197 ± 10	544 ± 36	864 ± 50
int8	222 ± 16	572 ± 37	862 ± 52

4 Conclusion

We quantized a deep neural network GAN model for a complex high energy physics use-case into a reduced precision format. As quantization tool we applied Neural Compressor from Intel to quantize a pre-trained model in float32 down to the int8 format. With the quantized int8 model we achieved a state-of-the-art inference time improvement of 3.0x compared to the not quantized float32 model on the latest Intel Ice Lake CPU. This results in a tremendous 120 000x speed-up over the presently employed Geant4 simulation which we aim to replace by modern generative models. Additionally, we achieved a 2.24x reduction in model memory size.

For the physics calorimeter use-case in this research, the highest possible accuracy is required. Therefore, we investigated the generated images very detailed with multiple physics metrics. The initial float32 model performed in most metrics very good. Furthermore, the quantized int8 model did not indicate a significant loss in accuracy compared to the initial float32 model.

The proposed reduced precision strategies for GAN models in this paper achieved encouraging results. These findings can help to reduce the required computing resources for future high energy physics simulations and for other generative model use-cases beyond. Therefore, as future work we want to improve our initial float32 model to reach a higher accuracy and repeat the reduced precision tests. For quantization we recommend to use accuracy driven quantization tools, as Neural Compressor in this study, in order to keep a high physics accuracy.

Acknowledgements. This work has been sponsored by the Wolfgang Gentner Programme of the German Federal Ministry of Education and Research.

References

1. Intel® neural compressor (2021). https://github.com/intel/neural-compressor
2. Agostinelli, S., et al.: GEANT4-a simulation toolkit. Nucl. Instrum. Meth. A **506**, 250–303 (2003). https://doi.org/10.1016/S0168-9002(03)01368-8
3. Elsen, E.: A roadmap for HEP software and computing R&D for the 2020s. Comput. Softw. Big Sci. **3**(1), 1–2 (2019). https://doi.org/10.1007/s41781-019-0031-6
4. Apollinari, G., et al.: High-luminosity large hadron collider (HL-LHC): technical design report V. 0.1 4/2017 (11 2017). https://doi.org/10.23731/CYRM-2017-004
5. Banner, R., Nahshan, Y., Hoffer, E., Soudry, D.: Post-training 4-bit quantization of convolution networks for rapid-deployment (2019)
6. Borji, A.: Pros and cons of GAN evaluation measures (2018)
7. Buhmann, E.: Getting high: high fidelity simulation of high granularity calorimeters with high speed (5 2020)
8. Feng Tian, Haihao Shen, J.G., Abidi, H.: Intel® lpot key takeaways (2021), https://www.intel.com/content/www/us/en/artificial-intelligence/posts/intel-low-precision-optimization-tool.html
9. Goodfellow, I.J., et al.: Generative adversarial networks (2014)
10. Gupta, R., Ranga, V.: Comparative study of different reduced precision techniques in deep neural network, pp. 123–136 (2021). https://doi.org/10.1007/978-981-15-8377-3_11
11. IEEE: IEEE standard for floating-point arithmetic. IEEE Std 754-2008, pp. 1–70 (2008)
12. Intel: oneAPI deep neural network library (oneDNN). https://github.com/oneapi-src/oneDNN
13. Itay Hubara, Yury Nahshan, Y.H., Banner, R.: Accurate post training quantization with small calibration sets (2021)
14. Jain, A., Bhattacharya, S., Masuda, M., Sharma, V., Wang, Y.: Efficient execution of quantized deep learning models: a compiler approach (2020)
15. Lu, L.: Dying relu and initialization: theory and numerical examples. Communications in Computational Physics **28**(5), 1671–1706 (2020). https://doi.org/10.4208/cicp.oa-2020-0165
16. Micikevicius, P., et al.: Mixed precision training (2017)

17. Nandakumar, S.R., Le Gallo, M., Piveteau, C., Joshi, V., Mariani, G., Boybat, I., et al.: Mixed-precision deep learning based on computational memory. Front. Neurosci. **14**, 406 (2020). https://doi.org/10.3389/fnins.2020.00406
18. Nasr, G.E., Badr, E., Joun, C.: Cross entropy error function in neural networks: forecasting gasoline demand. In: FLAIRS Conference (2002)
19. de Oliveira, L., Paganini, M., Nachman, B.: Learning particle physics by example: location-aware generative adversarial networks for physics synthesis. Comput. Softw. Big Sci. **1**(1), 1–24 (2017). https://doi.org/10.1007/s41781-017-0004-6
20. Osorio, J.: Evaluating mixed-precision arithmetic for 3D generative adversarial networks to simulate high energy physics detectors
21. Paganini, M., de Oliveira, L., Nachman, B.: CaloGAN: simulating 3D high energy particle showers in multilayer electromagnetic calorimeters with generative adversarial networks. Phys. Rev. D **97**(1), 014021 (2018). https://doi.org/10.1103/physrevd.97.014021
22. Pierini, M., Zhang, M.: CLIC Calorimeter 3D images: electron showers at fixed angle (2020). https://doi.org/10.5281/zenodo.3603122
23. PyTorch: Introduction to quantization on pyTorch (2020). https://pytorch.org/blog/introduction-to-quantization-on-pytorch/
24. Rehm, F., Vallecorsa, S., Borras, K., Krücker, D.: Validation of deep convolutional generative adversarial networks for high energy physics calorimeter simulations (2021)
25. Rehm, F., et al.: Reduced precision strategies for deep learning: a high energy physics generative adversarial network use case. In: Proceedings of the 10th International Conference on Pattern Recognition Applications and Methods (2021). https://doi.org/10.5220/0010245002510258
26. Swamidass, P.M. (ed.): MAPE (mean absolute percentage error). In: Swamidass, P.M. (ed.) Encyclopedia of Production and Manufacturing Management, p. 462. Springer, Boston (2000). https://doi.org/10.1007/1-4020-0612-8_580
27. TensorFlow Lite: Post training quantization. https://www.tensorflow.org/lite/performance/post_training_quantization
28. Vallecorsa, S., Carminati, F., Khattak, G.: 3D convolutional GAN for fast simulation. EPJ Web of Conferences **214**, 02010 (2019). https://doi.org/10.1051/epjconf/201921402010
29. Wang, N., Choi, J., Brand, D., Chen, C.Y., Gopalakrishnan, K.: Training deep neural networks with 8-bit floating point numbers (2018)
30. Wu, H.: Inference at reduced precision on GPUs (2019). https://developer.download.nvidia.com/video/gputechconf/gtc/2019/presentation/s9659-inference-at-reduced-precision-on-gpus.pdf
31. Wu, H., Judd, P., Zhang, X., Isaev, M., Micikevicius, P.: Integer quantization for deep learning inference: Principles and empirical evaluation (2020)

Similarity Constrained Conditional Generative Auto-encoder with Generalized Dilated Networks

Jan Niclas Reimann[1(✉)], Bhargav Bharat Shukla[1], Andreas Schwung[1], and Steven X. Ding[2]

[1] Automation Technology and Learning Systems,
South Westphalia University of Applied Sciences, Soest, Germany
{reimann.janniclas,schwung.andreas}@fh-swf.de
[2] Department of Automatic Control and Complex Systems, University of Duisburg-Essen,
Duisburg, Germany
steven.ding@uni-due.de

Abstract. Recent advancements in Generative Adversarial Networks have made it possible to generate plausible results. But what these models fail to learn is to disentangle different factors of variations to have a better control over the generated images. This makes generating images with specific features difficult. Typically, it is possible to do so, but with limited success, because models are prone to mix features from different classes together when generating images. For successful image generations with control over the content of the generated images, it is necessary that the model is interpretable conditionally disentangled. In fact, disentanglement is considered to be an important property of interpretability and is our focus point in this work. We introduce a novel idea to generate disentangled representations. In this approach, we add a Convolutional Encoder to the Conditional GAN structure and scale down the latent features to a vector. Labels are used at different levels in the Decoder with help of a separate network called Label-Scaler, which can be seen as a convolutional projection. Doing this, we drastically increase the disentanglement, which we visualize by performing traversals in both the feature latent space as well as in the conditional latent space. Additionally, we improve our intermediate results significantly by using Generalized Dilation Convolutions, increasing the reconstruction- and generation-performance of our framework without a change in disentanglement.

Keywords: Deep learning · Generative networks · Dilated convolutions · Constrained learning

1 Introduction

In recent years, image data generation in particular has been a keen area of interest for many researchers. Auto-Encoders (AE) and Generative Adversarial Nets (GANs [13]) are only two examples. One reason for the popularity is their ability to generalize on wide range of images to produce plausible results in various application domains [9, 20].

Deep Learning tasks in Computer Vision usually require lots of data to train on. While there is plenty of data available, the data is not always clean, labelled and ready

© Springer Nature Switzerland AG 2023
M. De Marsico et al. (Eds.): ICPRAM 2021/2022, LNCS 13822, pp. 23–51, 2023.
https://doi.org/10.1007/978-3-031-24538-1_2

to use for these algorithms. This makes pre-processing a very important step in the Machine-learning projects. GANs can help in reducing this pre-processing step by using the available labelled data to create synthetic samples. Another scenario can be in cases where the given dataset is imbalanced and the synthetic samples are used to balance out the classes. . An optimal solution is not always reached while training GANs, and there are very few sets of parameters where the GANs will converge, so GANs suffer from training instabilities and require a lot of tuning. One example of this instability is the well known mode collapse, which standard GANs are very sensible towards. Mode collapse occurs when the generator misses some modes in the data, resulting in a generator, which only generates samples in a limited range, in the worst case identical samples. Due to this problem, one area of interest in GANs and AEs is that of disentanglement in the latent space. One common defintion of disentanglement in this context is that disentanglement describes the ability of the trained GAN or AE to represent one feature of the reconstructed image or data as a single prior or noise value [16]. For example, if there are features such as color, size, and rotation represented by a vector, changing the rotation information should only change that angle of an object and not the other features. They also suggest that disentanglement leads to better feature representation and learning due to a more stable and reliable latent space.

We introduce two novel methods to solve the aforementioned shortfalls. First, we modify how the conditional information is passed to the generator framework. Here we add the conditional information at multiple places, using a label-scaler to dynamically adjust the label information to different target domains. To achieve disentanglement, we combine an Auto-Encoder and a GAN into one framework with additional losses to ensure cycle consistency. This helps to train the Encoder and Generator on adversarial losses in order for us to be able to use random noise to generate images. Without proper disentanglement and uncorrelating latent features random generations is not advised. We call this framework Similarity constrained AEGAN (SimAEGAN). The second novelty is using this approach as another use-case of our previously developed Generalized Dilation layer [5] (GDL) to significantly improve the performance of our approaches.

The core contributions of the paper can be summarized as follows:

- Achieving a disentangled feature representation when training Auto-Encoders.
- Conditional generation of data using interpretable attributes.
- Adding Gradient information as an additional evaluation metric when reconstructing data.
- A novel application of the Generalized Dilation Layer to prove its effectiveness not only in image classification and Remaining Useful Lifetime prediction, but also in image generation tasks.

The paper is organized as follows. Section 2 discusses the work related to our paper both for generative networks and dilated convolution kernels. Section 3 explains the basic idea of our proposed framework, how we train the framework is trained. Experiments are discussed in Sect. 4, including the corresponding results and analysis. Section 5 gives a brief summary of the paper and concludes our study.

2   Related Work

Our related work is split into two parts: the first part will describe related literature regarding the ideas of disentangled representation in the fields of AEs and GANs, while the second part will deal with related literature regarding the GDL.

2.1   Generative Networks

AEs were first introduced to explain the possibility of learning internal representations using back-propagation [39]. AEs are a form of unsupervised machine learning algorithm where the input is reconstructed after a set of compression and decompression operations, with the aim to reduce dimensionality of the data. With availability of computation capabilities, Convolutional Auto-Encoders (CAEs) can be scaled to multiple layers. As a result, it has seen a wide range of newer applications from the traditional use cases such as in Generative Modelling [29], Clustering [14], or anomaly detection [32].

CAEs, although good at tasks such as image restorations, completion, or denoising, fall short when it comes to sampling images from the latent space. To overcome this problem, methods to sample through the latent space with the help of Variational Auto-Encoders (VAE) [21] were introduced. The encoder is used as a recognition network to match the output with an arbitrarily chosen distribution such as Gaussian distribution, split into two outputs, mean and standard deviation, by using KL Divergence as a similarity metric. VAEs yielded good reconstructions and opened up the possibility to generate images, but images were typically blurry and dull. To overcome the shortcomings of VAEs, a modified of Auto-Encoder with a Discriminator network called Adversarial Auto-Encoders (AAE) was developed [29]. Here authors replace the KL Divergence with a adversarial training for the encoder. The use of discriminator in the training helps to reduce the blurriness in the generated and reconstructed images and makes them believable [2,36].

At the same time there has been a deep interest in the field of GANs [13], which demonstrated the power of adversarial training to generate synthetic images. Compared to the methods mentioned previously, GANs have the ability to generate sharp images which has allowed them to be popular in various computer vision tasks such as image generation [7,19,20], image translation [11,18,26,43] and image super resolution [25, 28] among the many.

Even though there has been a lot of progress, GANs are plagued by a few problems. First, the training dynamics of GANs are quite unstable. Second, they are hard to evaluate and third, the lacking ability to have control over the generated images or to have disentangled features. The Deep Convolutional GAN (DC-GAN) made it possible to train the GAN algorithm stably with help of Deep Convolutional Networks [37], which was further improved with the introduction of conditional GANs (cGANs) [33] with help of adding class information as conditional inputs to the training schematics. In a different version, the discriminator is not used for binary adversarial training, but rather a s a classifier [35]. It was also proposed to disentangle style and class information in an unsupervised manner using Mutual Information Maximization [6].

Another area of interest in GANs are loss criterions used to train the networks. Wasserstein GANs [1] use wasserstein distance to give a score to the images. Another group proposed the usage of Mean Squared Error as adversarial loss criterion for training GANs [30].

A possible extension that has got a lot of interest is combining both the power of disentanglement from VAEs and generating sharper images from GANs. Invertible GANs, for example attached two encoders to the GAN architecture, one for the latent output and other for class labels or attribute labels [36]. In 2017, two research groups happen to propose a similar idea of using an invertible AE structure while training the GAN [9, 10]. Ulyanov et al. [40] showed that it is entirely possible to achieve GAN like sampling from generator by only using AE structure. α-GANs extended this framework with help of a VAE and a discriminator to achieve a similar output [38].

These hybrid models are able to disentangle style and content and sometimes even features, but it is still not clear what is considered to be a disentangled representation. Higgins et al. [16] gives a general definition of disentangled representations. They claim that a vector is said to represent a disentangled output with respect to a particular decomposition of a symmetry group into subgroups, if it decomposes the group into independent subspaces affected only by the action of a single element to change the group by a single feature. They lay down the definition with three factors, first is Modularity, which states a single latent dimension should encode no more than single data generative factor. Second is Compactness, which states that each data generative factor is encoded by a single latent dimension but also suggest that it is possible that a data generative factor is shared by two or more latent dimensions. Third is Explicitness, which is defined by the ability that all data generative factors can be decoded using a linear transformation.

Since there is no single loss value that signifies the diversity or quality of generated images Borji [3] carried out a detailed survey identifying pros and cons of different GAN evaluation metrics. Although there are many available metrics to evaluate GANs, they somehow fall short of being completely optimal. For example, Inception Score and Fréchet Inception Distance, two of the most widely used methods, use a pre-trained network to calculate these measures which may not adapt well to other domains (faces, digits).

2.2 Dilated Convolution Kernels

Since the main idea why we want to use dilated convolution kernels is to get information of a larger context, while preserving the fine details (i.e. edge information), this scenario is a fitting new application for the previously published GDLs [5]. Most of the related work still holds, so we will briefly mention key concepts in the context of image generation. Similar to image classification, our receptive field is two dimensional and should also capture the spatial information in a certain range. Since not all pixels are important, the concept of dilation allows for a higher receptive field size, without increasing the number of parameters. Optimizing integer values is a problem on its own, so there exist different approaches how to implement ideas similar to the dilation into convolutional kernels. Layers with a variable dilation factor typically realize this by utilizing bilinear transformations of the weight matrix, while keeping the structure of the dilation fixed [8, 15, 41]. Compared to our previous work, we use the GDN not to focus on specific regions in the input kernel, but to have a dynamic adjustment of local gradient computation, resulting in a small change described in the Section. 3.2. Here, the gradient is the gradient over the input image, not the gradient the parameters are trained with. Using the gradient information of the input image as well as the difference in terms of raw pixel data seems redundant on the first glance, but should put a stronger emphasis on sharp edges at the correct locations. To the best of the authors knowledge using the gradients of the input image as well as the pixel difference is a novel approach.

3 Fundamentals

In this Section we will describe the various approaches in detail. First, we will describe the basic model of the SimAEGAN. Second, we will introduce the ideas of the Label Scaler extension, before explaining how this is related to the usage of GDLs. For all approaches we explain all the hyperparameters, the loss calculations and the general training setup.

3.1 SimAEGAN

To lay the basis for our experiments, we use an architecture which is similar to the ones used in [22, 23, 38, 42]. The main components of this newly proposed architecture are an Encoder E, Decoder/Generator G, Disc_{img}, Disc_{feat} and a Label Scaler L_s. Figure 1 shows the proposed architectures in detail.

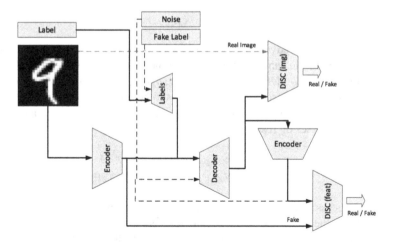

Fig. 1. Overview of the proposed SimAEGAN.

Training Framework. The training begins with calculating the loss in reconstruction between the input images x and reconstructed images x'. The Encoder E is a CNN trained to produce latent features z in a range from -1 to 1. We restrict the latent code to a smaller latent feature space to force E to learn meaningful features. The Decoder and/or Generator G accepts these latent features and the conditional information (class labels/attribute labels) y as inputs and reconstructs the image as shown in Eq. (1). The conditional information is one-hot encoded, if applicable, and are passed through the Label Scaler L_s to up-sample the labels to different feature sizes. In general, both L_s and the G share the losses and weight updates. The architecture of G with L_s is explained in Subsect. 3.1. Here λ_{MAE} is the loss weight for back-propagating the losses for Auto-Encoder phase.

$$\mathcal{L}_{MAE} = \lambda_{MAE}\text{MAE}(x', x) \tag{1}$$

Next, the noise $\tilde{z}$ and generated labels $\tilde{y}$ are sampled from a prior distribution. The noise is either a Gaussian distribution or Uniform distribution in range from -1 to 1 and the labels are selected randomly with an equal distribution for each class individually. In cases of attribute labels, the generated labels are directly sampled from the training labels y. We assume that since there are multiple combinations for all the attributes, they are unlikely to over-fit to the conditional information as each combination of attributes would require large enough sample size to over-fit based on samples.

The sampled noise and the generated labels are passed through G to produce generated samples $\tilde{x}$. These generated images are then passed through E to recover the noise back from E. This helps to make E-Decoder structure invertable by using an idea similar to cycle consistency [44]. The difference between the noise $\tilde{z}$ and reconstructed noise z' is calculated using mean squared error for the samples as shown in Eq. (2). Here λ_{MSE} is loss weight for back-propagating the losses to E.

$$\mathcal{L}_{MSE} = \lambda_{MSE}\text{MSE}(z', \tilde{z}) \tag{2}$$

Then we start with the adversarial training for E. In this case, E is trained to match the outputs(z, z') with the sampled noise $\tilde{z}$. Here the Disc_{feat} is trained based on the probability of the Disc_{feat} identifying the outputs as real. This trains E adversarially to match the sampled noise distribution (Eq. (3). It is to be noted that only E is trained in this part and the weight updates to the generator are ignored. Here $\lambda_{E_{adv1}}$ and $\lambda_{E_{adv2}}$ are the weights for losses on respective values.

$$\begin{aligned}\mathcal{L}_{E_{adv}} = &- \lambda_{E_{adv1}}\mathbf{E}_{x\sim p_d}\log(D_{feat}(E(x))) \\ &- \lambda_{E_{adv2}}\mathbf{E}_{\tilde{z}\sim p_d}\log(D_{feat}(E(G(\tilde{z}|\tilde{y}))))\end{aligned} \tag{3}$$

A similar process is done again when training G. G is trained on the probability of Disc_{img} such that the generated images $\tilde{x}$ and reconstructed images x' are detected as real images (Eq. (4)). Similar to previous step, only G is trained and the weight updates to E are ignored. Here $\lambda_{G_{adv1}}$ and $\lambda_{G_{adv2}}$ are the weights for losses on respective values.

$$\begin{aligned}\mathcal{L}_{G_{adv}} = &- \lambda_{G_{adv1}}\mathbf{E}_{x\sim p_d}\log(D_{img}(G(E(x)|y))) \\ &- \lambda_{G_{adv2}}\mathbf{E}_{\tilde{z}\sim p_d}\log(D_{img}(G(\tilde{z}|\tilde{y})))\end{aligned} \tag{4}$$

Then Disc_{img} is trained using all three sets of images. The task of the Disc_{img} is to correctly identify input images as real and to detect both generated and reconstructed images as fake. To overcome the bias from having more generated images, we add extra input images from the real set (Eq. (5)). It is to be noted that loss updates are only made for Disc_{img} and loss for Encoder and Decoder are ignored. Here $\lambda_{D_{img}}$ is the weight value for the loss on the reconstructed images.

$$\begin{aligned}
\mathcal{L}_{D_{img}} = &- \mathbf{E}_{x \sim p_d} \log(D_{img}(x)) \\
&- \lambda_{D_{img}} \mathbf{E}_{z \sim p_d} \log(1 - D_{img}(x')) \\
&- \mathbf{E}_{\tilde{z} \sim p_d} \log(1 - D_{img}(\tilde{x}))
\end{aligned} \tag{5}$$

Similarly the Disc_{feat} is also trained on three sets of features, where noise is the real input and the reconstructed noise and latent features are marked as generated (Eq. (6)). The Disc_{feat} also is trained on equal number of real and generated inputs. Only Disc_{feat} is updated with the loss and loss for Encoder and Decoder is ignored. Here $\lambda_{D_{feat}}$ is the weight value for the loss on the reconstructed noise.

$$\begin{aligned}
\mathcal{L}_{D_{feat}} = &- \mathbf{E}_{x \sim p_d} \log(D_{feat}(\tilde{z})) \\
&- \lambda_{D_{feat}} \mathbf{E}_{z \sim p_d} \log(1 - D_{feat}(z')) \\
&- \mathbf{E}_{\tilde{z} \sim p_d} \log(1 - D_{feat}(z))
\end{aligned} \tag{6}$$

The convergence criterion for our training is to train for a certain number of epochs. We refrain from early-stopping method as GANs are a two player game, it is difficult to judge when the training has converged due to the previously discussed problems of not having a proper measurement of the quality of images.

Label Scaling in Detail. The traditional method of concatenating the label information directly to the noise may result in the network completely or partially ignoring the label information. This leads to the cGANs producing samples of unequal quality or possibility of class mixing for some features. To overcome this problem, one can either use an embedding layer to convert the labels into a different information domain and then up-sample to match the image size, as mentioned in [12], or by using an auxiliary classifier to the Discriminator so that the Discriminator has two tasks, the adversarial one and to classify the class of the image [35]. We build upon the first idea but also modify the idea which has been explained below.

Fig. 2. Overview of one Conv-Block in the generator.

Figure 3 shows the architecture of the modified conditional generator in detail. The top row consists of a Label Scaler network which is used for scaling the labels to different feature sizes and latent spaces. The bottom row is the generator network. Up-sampling + Convolutional Layer are preferred to increase the feature sizes instead of

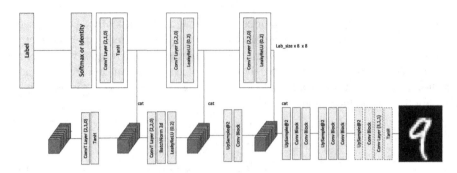

Fig. 3. Modified generator with label scaler.

traditional Transposed Convolutional layers as it is shown that the later method leads to checkerboard effects in the generated images [34].

In the Label Scaler, the class labels or the attribute vector is soft one-hot encoded that is, instead of having one-hot label for class three in total of three classes [0., 0., 1.], small noise is added to the zeros and the same is subtracted from the one in the vector to result [0.01, 0.01, 0.98]. This is done in-order to add some non-linearity in the labels and convert the sparse input into dense input, similar to labels smoothing in typical GAN applications. If the labels are not added with noise then we use a Softmax layer to impose non-linearity. Then the labels pass through the multiple Transposed Convolution layers and concatenated at respective sizes.

The first two layers of the generator network consists of Transposed Convolutional layers to increase the feature sizes upto 4×4. The rest of the generator network is made up of Up-sampling and multiple ConvBlocks or ResBlocks as shown in Fig. 2. The ConvBlocks are convolutional layers followed by Batch-Normalization layer followed by a non linear activation function such as Leaky ReLU with a negative slope of 0.2. The convolutional layer are used to increase or decrease the number of features in the network while keeping the latent feature size constant by choosing the padding accordingly.

3.2 Dilation

As an extension to the above mentioned architectures we will use Generalized Dilated Networks (GDNs) [4,5] with the goal of reducing artifacts and generating sharper images. Here, we will briefly introduce the main concept of GDNs. In a Generalized Dilation Layer (GDL), an additional matrix, the masking matrix $\tilde{\Psi}$, is introduced. This matrix has the same shape as the convolution kernel and is elementwise multiplied with the weight matrix before computing the strided convolution. $\tilde{\Psi}$ is not only trained on the classification loss $L_s(\omega, \tilde{\Psi})$, but also by the barrier function loss $L_b(\tilde{\Psi})$. As barrier functions we use the same barrier functions as in [5]:

$$b_c(x) = b_r(x) = \max\left(\left(e^{\alpha_1 \cdot (x-p)} \cdot \alpha_2 \cdot (x-p)\right), \alpha_3\right), \tag{7}$$

$$b_a(x) = \max\left(\left(e^{\alpha_1 \cdot (x^2-p^2)} \cdot \alpha_2 \cdot (x^2-p^2)\right), \alpha_3\right), \tag{8}$$

$$\nabla_{\tilde{\Psi}} L_b = b_c(\tilde{\Psi}) + b_r(\tilde{\Psi}) + b_a(\tilde{\Psi}), \tag{9}$$

where b_c, b_r, b_a represent the barrier functions over the columns, the rows, and over all entries, respectively. Using Eq. 8 to Eq. 10, the local loss is defined as

$$\nabla_{\tilde{\Psi}_{ij}} = \frac{\partial L_s(\omega, \tilde{\Psi})}{\partial \tilde{\Psi}_{ij}} + \nabla_{\tilde{\Psi}_{ij}} L_b. \tag{10}$$

Using these barrier functions, we force the dilation mask to result in sparse matrices. This should put a stronger focus on generating sharp edges and reducing blurriness.

Due to the use of the tanh-function, we also changed the parameter initialization and barrier functions accordingly. The barrier functions do not compute the barrier loss based on $\sigma(\tilde{\Psi})$, but on $|\tanh(\tilde{\Psi})|$.

Before applying the GDLs on this approach, we used manually generated kernels for derivatives in x- and y-direction to not only map the output pixels to the input pixels (i.e. reconstruction loss), but also map the output gradient to the input gradient (gradient loss over the pixel data). This was not as beneficial as expected, thus we made this step learnable so the network can adapt and find a more suitable mask on its own.

4 Experiments

In this section we will introduce our experiments and ablation studies. First, we show a proof of concept on the MNIST [24] and a customized version of the Dsprites dataset [31]. Afterwards, we show a thorough study on the CelebA dataset [27], where we also compare our generated and reconstructed images, as well as feature traversals of said images. Traversals are step-wise changes in a constant vector, where we only change one feature at a time, from the minimum value to the maximum value. These traversals are done on the feature fector, as well as on the conditional information. Here, we also give a brief summary of the results below.

Table 1. Shared Hyper-parameters used for all the experiments.

Parameter	Value	Parameter	Value
Noise Distribution	Uniform	Image Size	32/64
Lr_E	0.0001	Lr_G	0.0001
$Lr_{D_{img}}$	0.0001	$Lr_{D_{feat}}$	0.0001
β_1, β_2	(0.50, 0.999)	Batch Size	100
λ_{MAE}	1.0	λ_{MSE}	0.5
$\lambda_{E_{adv1}}$	0.0001	$\lambda_{E_{adv2}}$	0.0001

In Table 1 we show all hyper-parameter shared between all following results. The Image Size depends on the dataset, where we rescale MNIST dataset to 32×32 and cSprites and CelebA to 64×64. We transform the input images in range of -1 to 1 and do not use any further image augmentation techniques. The optimizer for all the networks is Adam. We train for 50 epochs and reduce the learning rate by 90% every 20 epochs. The details of network architectures have been attached in Tables 2 to 8 (appendix).

Considering the task at hand, we set the hypothesis that reducing the latent space of the Decoder forces it to learn the features efficiently, while also helping in the case of feature-disentanglement. We set a baseline model (without GDLs) and test on customized Dsprites and MNIST dataset. The baseline model is able to disentangle the features, but we will show that it suffers from drawbacks for CelebA dataset which are explained in Sect. 1. The model wcan be seen as a modified AE-GAN structure, combining idea from both DC-GAN, and AAE.

For CelebA dataset, we first train our model only on the previously mentioned training setup in Subsect. 3.1. For the runs with the original model, we see that the reconstructed and generated images contains artifacts around the face regions and are blurry. Training longer also does not solve the issue. Moreover, the reconstructed images show a low resemblance to the original image. It is to be noted however, that manipulations in the latent space are successful i.e. the original image attribute male can be change to female to obtain an image of a female with the same features.

To overcome the effect of artifacts and overall blurriness of the reconstructed or generated images, we use the Generalized Dilation Layer as a similarity constraint between the original and reconstructed images. The motivation behind this is that two similar images should generate similar features and therefore the loss between the two sets of images should be minimized. Here we use three GDLs, where each layer hase three input channels and one output channel. Field sizes are fixed to 11, whereas the kernel sizes are set to 5, 7, and 9. The optimizer for the generalized Dilation layer is Adam with the default learning rate of 0.001.

First we train the algorithm on our base training setup without using the GDLs, we only enable the similarity loss and comparison the generated features from both the original and reconstructed images with Mean Absolute Error of the two set of features for the last 20 epochs. The Dilation layer, as well as the Decoder and the Encoder, are trained during the backward pass of the loss. We modify the original Dilation layer [5] to make it possible to have positive and negative weights, indicating a pseudo-gradient calculation. We see that the generated and reconstructed images show improved reconstruction quality (Figs. 13, 14) and are also less blurry when compared to runs without the use of dilation layer as feature similarity constraint (Figs. 9, 10).

4.1 Custom dSprites

Here, we show our results on the custom dSprites dataset. We extended the original dSprites dataset [17] by adding different shapes (Square, Circle, Ellipse, Trapezoid), as well as color information (Red, Blue, Green) and call the resulting dataset cSprites (custom dSprites). We select the color and shape information as our conditional values, whereas the other features (x-position, y-position, size,...) are randomly sampled. In total we sample 1680 images as our dataset. The images are split randomly with 70 percent for training and the remainder for testing data. We use the same hyperparameters as described in Table 1. Detailed network architectures are shown in Tables 2, 7, 4 and 5 for E, G, Disc_{feat} and Disc_{img}.

Since our generator supports two different inputs (the conditional information as well as the random noise), we choose a subset of attributes as our conditional information and leave it to our network to generate the other information by itself. Since we selected the color and shape information as our conditional information, we sample each of these values individually as one-hot encoded labels. All the other information (e.g. x and y position and size) are randomly sampled since they should be a result of our encoder.

Figure 4 shows multiple generated images, where each column affects only single conditional information.

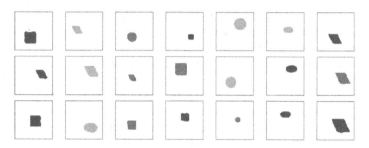

Fig. 4. Generated cSprites images. Each column represents three generations with the same conditional information: blue, green, red, square, circle, ellipse, and trapezoid (Color figure Online)

Obviously the generation of such simple shapes works, but the conditional information is interpretable and for all intents and purposes completely disentangled. We can select the conditional information, randomly sample the remaining features and generate a corresponding sprite. Since our remaining features should capture position and scale, we now keep the conditional information static and traverse through each latent code individually. The results are depicted in Fig. 5.

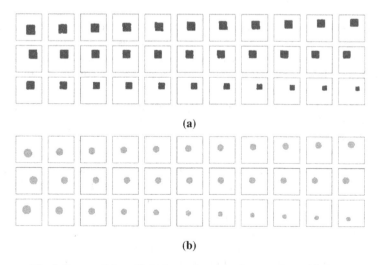

Fig. 5. Traversal through the latent features of two generated images.

Here, we observe the same results as when analyzing the conditional inputs: our encoder outputs are also for all intents and purposes disentangled and interpretable. We know, that the latent features from our encoder represent y-position, x-position, and the scale of the sprite (row one, two and three in Fig. 5 respectively), exactly the three features we left for the encoder to detect. Overall the framework is definitely working in producing a disentangled and interpretable latent space.

4.2 MNIST

Similar to the cSprites-results, we use the MNIST dataset as a proof of concept. The MNIST dataset is arguably more complex than the cSprites dataset due to its wider spread of information in the input image. Shapes are more complex than basic sprites and the object fills out a much bigger area of the input image, making capturing information over a wider region much more important, while also introducing different styles of the same object.

The Dataset consists of 70000 images with 60000 images as training data and the remaining 10000 as testing data. The dataset consists of 10 classes of grey scaled digits from 0 to 9. Tables 3, 4, 6, and 8 show the detailed architectures for E, G, $Disc_{feat}$ and $Disc_{img}$ respectively. The hyper-parameters are also left unchanged as per Table 1.

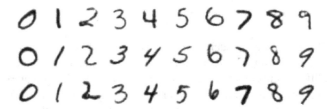

Fig. 6. Conditionally generated images for the MNIST dataset. Each column represents three generations with the same conditional information.

The generated images look realistic, while being conditionally generated (Fig. 6). Obviously the conditional information is working, whereas the other (random) parameters have a different impact on the generated images: they combine position, thickness, sizes, rotation, different styles and various other features. The conditional information is still highly disentangled, interpretable and reliable. Since our architecture consists of two parts (GAN and AE), Fig. 7 shows the reconstruction performance on images from the testing dataset.

The reconstruction performance is also as expected. All reconstructions share the most common features of the input image, while still lacking the minor details, such as small twirls or faded out ends. Typically, using parts of AEs, especially non-variational AE, to generate images results in way worse generations, due to the correlation/entanglement of the latent features. By introducing the various losses mentioned in Sect. 3.1 we are able to achieve a disentangled latent representation of our encoder outputs. To analyze this, we computed the Encoder outputs for each image in the testing dataset and separated them according to their conditional information (their class) with and without our additional losses. Figure 8 shows the distribution of one of these latent features.

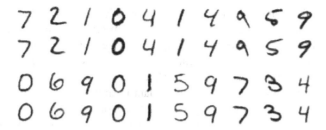

Fig. 7. Reconstructions from samples of the testing MNIST datset. Odd rows show the original image, even rows the reconstructed image.

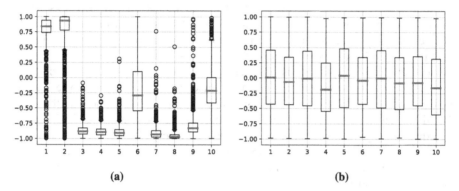

Fig. 8. Distributions of one latent feature (one output of E), separated by classes. (a) training without our additional losses (b) SimAEGAN.

We see, that the latent features in the original AE section are highly condensed to small areas in the possible output space (Fig. 8a). However, if we train the same network with our introduced losses enabled (Fig. 8b), it results in a "proper" distribution, closely resembling the prior distribution which was used in the generator training (uniform from $[-1 \ldots +1]$). This output distribution explains why we are able to exchange our encoder part of the network with random noise to successfully generate images. Using our proposed losses we do not need to analyze our encoder outputs in retrospect. We do not need to analyze the correlation of encoder outputs before generating images to make sure we are sampling in the correct subspace.

4.3   CelebA

The CelebA dataset [27] is a common dataset for generative tasks. It contains 202,599 images of 10,177 celebrities with 40 boolean features each. For easier visualization we select a small subset of features based on their distribution in the dataset and their complexity: we want the features to be evenly distributed (meaning almost an almost 50:50 split) and mostly to be pretty simple to see the changes visually. The selected attributes are Bangs, Black_Hair, Eyeglasses, Heavy_Makeup, Male, Receding_Hairline, Smiling, Straight_Hair, Wavy_Hair and Young.

We selected these attributes over the other not only because of their distribution, but also because of the nature of these attributes themselves. Most selected attributes only correspond to local changes in the image and can easily be observed (i.e. Bangs, Black_Hair, Eyeglasses, Receding_Harline, Smiling, and both of the hair-attributes), while other attributes (i.e. Heavy_Makeup, Male, and Young) have a bigger and more complex impact on the images. The training data is split to contain around 1,62,800 images and 20,000 images for testing. Detailed network architectures are shown in Tables 2, 7, 4 and 5 for the E, G, Disc_{feat} and Disc_{img}. The hyper-parameters are also left unchanged as per Table 1.

Without GDLs. To compare the effect of the additional dilation networks on the whole architecture, we will show results without GDLs first. Here, we will focus on three different aspects: image reconstruction, image generation and disentanglement/interpretabiliy of latent features. These aspects were also used in the previous two datasets, but the aspect of disentanglement and interpretability requires additional explanation of what we understand as disentanglement when it comes to features of the CelebA dataset. Whenever we change one conditional feature, the only thing that should happen (in a case of perfectly disentangled features) is a change in the output according to the semantic meaning of this individual feature. If we add eyeglasses to an image (or remove eyeglasses from an image) it should only affect the region around the eyes, but not the hair, the gender, the background etc. When we change the more complex features, we also expect a bigger change of the overall image. These more complex features were missing as conditional information in cSprites and MNIST.

Analog to cSprites and MNIST, we will start with the reconstruction performance. We sample various images from the testing dataset and show the reconstructions of four testing images in Fig. 10. The first row represents the original image, the second row represents the reconstruction of our architecture.

(a)   (b)   (c)   (d)

Fig. 9. Reconstructed images from the CelebA dataset without additional GDLs.

In all reconstructions we see a significant difference in quality. Reconstructed images have considerable background artifacts, but also artifacts in the reconstructed faces themselves (e.g. Figs. 9c and 9c), with artifacts being big colored stripes and inconsistent edges. Based on the reconstructions we assume that the decoder effectively only supports the information of the before mentioned more complex conditional features, features which affect the whole image and thus have a big impact on the reconstruction performance. HairColor, HairStyle, and Smiling are also somewhat consistent, but the network was unable to correctly restore the remaining attributes. A similar performance can also be seen in the generation of images, where we can observe similar artifacts and inconsistencies (Fig. 10). To generate random images, the feature vector (which would otherwise be the encoder output) is sampled from $\mathbb{R}$ with a uniform distribution between -1 and $+1$ (uniform distribution to mimic the tanh of E). The conditional information is randomly selected to be 0 or 1 for each attribute individually.

We can see that the aforementioned artifacts, especially the ones in the background,

Fig. 10. Conditionally generated images without additional GDLs.

seem to be a trained bias in the network. We found out that this bias is not from the conditional information, but rather from the feature information (see Fig. 20). Many of the other common artifacts also seem to correspond to these features, indicating either that the generator is still lacking training, or that the encoder features (even though they are properly distributed) are slightly correlated, since artifacts (such as the artifacts in Figs. 9c and 9c) are not constantly visible in all images, but rather only for some of them.

Since we know which conditional attribute represents which high level information in the resulting image, we encoded images from the testing dataset and manually changed the latent information in multiple steps. Figure 11 shows the end-points while traversing through the attribute vector of an encoded image from the testing dataset.

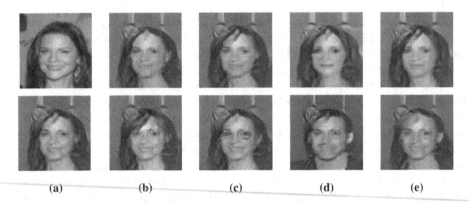

Fig. 11. Traversals of conditionally reconstructed images without additional GDLs. (a) original/reconstructed image (b) attribute *bangs* (c) attribute *eyeglasses* (d) attribute *female/male* (e) attribute *smile*.

These attribute traversals match the previous assumptions. The network is not always able to capture the lower level facial attributes and mostly reacts to changing the higher level attributes, such as gender. We also see, that various disturbances, among that the artifacts, are not changed by the lower level attributes. Even though the conditional attributes can not be sufficiently generated, they only impact their actual region of interest, e.g. the hair strain on the forehead is visible in all images which do not deal with the *bangs* attribute. This behavior is also consistent when we are not traversing with encoded features, but with features from a generated image (see Fig. 12).

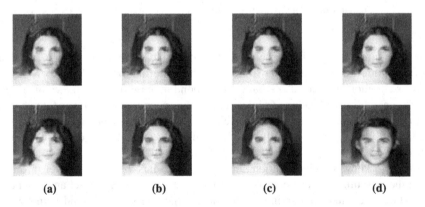

Fig. 12. Traversals of conditionally generated images without additional GDLs. (a) attribute *bangs* (b) attribute *eyeglasses* (c) attribute *smile* (d) attribute *female/male*.

The approach, which was shown in the case of the two toy datasets/proof of concept datasets cSprites and MNIST, is also working on the much more complex CelebA dataset. Both the reconstruction and generation performance is severely suffering from the increased complexity and restrictions to achieve a disentangled representation. Since we are only evaluating the reconstructed and generated images on a very small scale (pixel-wise optimization from the AE part) and the highest possible scale (using the discriminator from the GAN part) an approach to quantify the performance on an intermediate scale was the next approach. One problem in the reconstructed and generated images is that of the big artifacts. Small changes in each of the RGB channels of the image might only have a small impact with regards to the loss, but the resulting color when ploting the image is significantly different for the human perception. Since we assume that the discriminator focuses on general facial attributes instead of artifacts in the background (which is where they typically can be found), the plan is to introduce the GDNs as a seperate evaluation part parallel to D_{img}.

With GDNs. For the following experiments we added the GDN as an evaluation layer parallel to D_{img}. The rest of the architecture, including all hyperparameters is kept the same to have the best possible comparison and best possible traceback to the effect of the GDN.

When comparing the reconstructions of the network without the GDN (Fig. 9) and with the GDN (Fig. 9) it becomes clear that there is a significant performance increase. The reconstruction performance is now much better than before. Artifacts are reduced,

(a)   (b)   (c)   (d)

Fig. 13. Reconstructed images from the CelebA dataset with additional GDLs.

images are less blurry, have more details, and have sharper edges. Images look mostly as real images, even though they are missing certain features (pose information, specifics in the background, azimouth etc.). We are training part of the network as an AE, but the GAN part of training does not have access to the input image, thus leading the network to different local features. Overall the reconstructions just seem to be a much clearer and accurate representation than the reconstructions without GDNs, elevating most of the problems of the framework without GDNs.

As before, the reconstruction performance gives us a good idea of what to expect in the generated images, although the performance increase is even more recognizable in the generated images (Fig. 14). In the previous approach, nearly all generated images had not only big artifacts in the background, but also in the regions around their hair and eyes. Most of the artifacts in the image are gone, and the artifacts left are typically

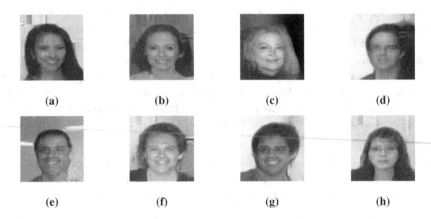

(a)   (b)   (c)   (d)

(e)   (f)   (g)   (h)

Fig. 14. Conditionally generated images with additional GDLs.

scattered around the corner of the image and do not directly affect the generated faces. One attribute, which previously was very rarely working, was the attribute of the eyeglasses. Without the GDN, eyeglasses were generated as dark patches around the eyes, not representing eyeglasses. Now, they are actually generated with temples and small glass patches (e.g. Fig. 14g). The big gain in performance by using GDNs can also be seen in the attribute traversals of a reconstructed image, especially for the attribute eyeglasses (Fig. 16b). The source image for this reconstruction is the same as in Fig. 11.

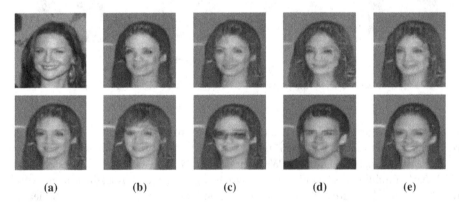

(a)   (b)   (c)   (d)   (e)

Fig. 15. Traversals of conditionally reconstructed images with additional GDLs. (a) original/reconstructed image (b) attribute *bangs* (c) attribute *eyeglasses* (d) attribute *female/male* (e) attribute *smile*.

The disentangled latent representation did not seem to suffer by introducing the GDNs to the network (this can also be seen in Figs. 21 and 20, appendix), while significantly improving the visual clarity of generated as well as reconstructed images. Changes of local attributes still only affect the local region of importance, while changes in more complex conditional information change the overall image, while keeping the rest of the features unchanged. This can be seen in the bangs and smile attribute (see Fig. 15a) when changing the gender from female to male (Fig. 16d).

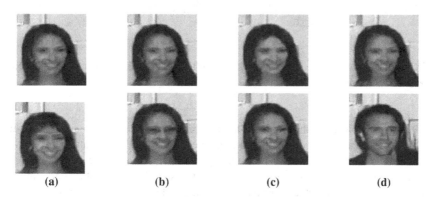

(a)   **(b)**   **(c)**   **(d)**

Fig. 16. Traversals of conditionally reconstructed images with additional GDLs. (a) attribute *bangs* (b) attribute *eyeglasses* (c) attribute *smile* (d) attribute *female/male*.

To more accurately represent the impact of the latent features and to make sure that we still avoided mode collapse, we sampled a random conditional vector and generated images with random feature vectors. Figure 17 shows that we can generate a variety of different faces using identical conditional attributes.

Fig. 17. Conditionally generated images with identical conditional attributes: no bangs, no black hair, eyeglasses, no makeup, male, receding hairline, not smiling, straight hair, no wavy hair and young.

In a visual comparison of our model with a VAE-GAN [22], we can clearly see that our model works better in disentangling different attributes and does not result in unwanted change in the reconstructed image or generated image. This is not the case with VAE-GANs (Fig. 18), where it is clearly seen that changing attributes result in additional changes.

(a)	(b)	(c)	(d)

Fig. 18. Attribute traversals for VAE-GAN [22] for different attributes. (a) input image, (b) Bald, (c) Eyeglasses, (d) Moustache.

We can see, that even though VAE-architectures are widely considered to be disentangled (or at least more disentangled), changes in the conditional attributes heavily influence other global features. When making the "person" in the upper image bald, it also impacts the gender and age information, the same happens when changing the feature eyeglasses or moustache. Even for an image where the features might be more fitting (a male person), we see similar results. When making the person bald, he automatically gets older and gets a beard. When adding glasses, a beard is also added.

Randomly conditionally generated images using VAE-GANs are able to reconstruct and generate the images with a better overall visual quality, which we believe is partly because of zooming into the faces and making the background information almost irrelevant, but are significantly worse in terms of disentanglement.

5   Conclusion

In this work we introduced a novel architecture, which we termed SimAEGAN. By cycling through the different domains in our network we achieve proper distributions in our latent space, which significantly improves the generation capabilities of AEGANs. We also introduced the labels/attributes at multiple locations in our decoder, each projected to a self-learned representation, since different features might be relevant at different stages during generation and reconstruction. This proved to be highly effective in terms of disentanglement of our latent features, as shown in our feature traversals (Fig. 15and 17). Finally, the addition GDNs reduced not only the blurriness in our images without loosing any disentanglement performance, but also removed most of the unwanted artifacts.

In future research we plan to tune our network for even sharper reconstructions and generations, without sacrificing disentanglement performance. We also want to test the effect of incorporating the generated images into different applications, e.g. use conditionally generated examples to balance out datasets, increasing the generalization performance when only limited data is available.

Appendix

Network Architectures

Table 2. Encoder for CelebA and cSprites.

Layer	Input size (C × H × W)	Filters units (width)	K, S, P	bias
Input	3 × 64 × 64	–	–	–
Conv + BN + LReLU	3 × 64 × 64	64	5,1,0	False
Conv + BN + LReLU	64 × 60 × 60	128	5,1,0	False
Max-Pool 2d	128 × 56 × 56	–	2,2,0	–
Conv + BN + LReLU	128 × 28 × 28	128	5,1,0	False
Conv + BN + LReLU	128 × 24 × 24	256	5,1,0	False
Max-Pool 2d	256 × 20 × 20	–	2,2,0	–
Conv + BN + LReLU	256 × 10 × 10	256	5,1,0	False
Conv + BN + LReLU	256 × 6 × 6	256	5,1,0	False
Conv + Tanh	256 × 2 × 2	**feat**	2,1,0	True
Flatten	**feat**	–	–	–

Table 3. Encoder for MNIST.

Layer	Input size (C × H × W)	Filters units (width)	K, S, P	bias
Input	1 × 32 × 32	–	–	–
Conv + BN + LReLU	1 × 32 × 32	64	4,2,1	False
Conv + BN + LReLU	64 × 16 × 16	128	4,2,1	False
Conv + BN + LReLU	128 × 8 × 8	128	3,1,1	False
Conv + BN + LReLU	128 × 8 × 8	256	4,2,1	False
Conv + BN + LReLU	256 × 4 × 4	256	3,1,1	False
Conv + Tanh	256 × 4 × 4	**feat**	4,1,0	True
Flatten	**feat**	–	–	–

The tables explain the network architectures used for various datasets for the experiments. Here **(K,S,P)** refers to Kernel, Stride and Padding respectively.

Here in Table 4, linear feats refer to the feature outputs of the encoder or noise reshaped to $(N, -1)$ where N is the number of samples and -1 is the flattened output.

Table 4. Feature Discriminator (Disc_{feat}).

Layer	Input size (C × H × W)	Filters units (width)	K, S, P	bias
Input	**linear Feats**	–	–	–
Linear + LReLU	**linear Feats**	64	–	True
Dropout(p = 0.5)	64	–	–	–
Linear + Sigmoid	64	1	–	True
Output	1	–	–	–

Table 5. Discriminator (Disc_{img}) for CelebA and cSprites.

Layer	Input size (C × H × W)	Filters units (width)	K, S, P	bias
label scaler				
Linear + Tanh	**Label Size**	256	–	–
Reshape	1 × 16 × 16	–	–	–
Discriminator Network				
Input	**Ins** × 64 × 64	–	–	–
Conv + LReLU	**Ins** × 64 × 64	48	4,2,1	False
Conv + LReLU	48 × 32 × 32	48	4,2,1	False
Conv + LReLU	48 + 1 × 16 × 16	96	4,2,1	False
Conv +LReLU	96 × 8 × 8	96	4,2,1	False
Conv	96 × 4 × 4	192	4,2,1	False
LReLU	192 × 2 × 2	–	–	–
Conv + Sigmoid	192 × 2 × 2	1	2,1,0	True
Flatten	1 × 1 × 1	–	–	–
Output	1	–	–	–

Table 6. Discriminator (Disc_{img}) for MNIST.

Layer	Input size (C × H × W)	Filters units (width)	K, S, P	bias
label scaler				
Linear + Tanh	**Label Size**	256	–	–
Reshape	1 × 16 × 16	–	–	–
Discriminator Network				
Input	**Ins** × 64 × 64	-	-	-
Conv + LReLU	**Ins** × 64 × 64	32	4,2,1	False
Conv + BN + LReLU	32 × 32 × 32	32	4,2,1	False
Conv + BN + LReLU	32 + 1 × 16 × 16	64	4,2,1	False
Conv + BN + LReLU	64 × 8 × 8	64	4,2,1	False
Conv	64 × 4 × 4	128	4,2,1	False
BN + LReLU	128 × 2 × 2	–	–	–
Conv + Sigmoid	128 × 2 × 2	1	2,1,0	True
Flatten	1 × 1 × 1	–	–	–
Output	1	–	–	–

Table 7. Generator for CelebA and cSprites.

Layer	Input size (C × H × W)	Filters units (width)	K, S, P	bias
label scaler				
Lab_{in}	**Label size** × 1 × 1	–	–	–
Lab_2 ConvT + Tanh	**Label Size** × 1 × 1	**Label Size**	2,1,0	True
Lab_4 ConvT + LReLU	**Label Size** × 2 × 2	**Label Size**	2,2,0	True
Lab_8 ConvT + LReLU	**Label Size** × 4 × 4	**Label Size**	2,2,0	True
Main network				
Input	**Ins** + lab_{in} × 1 × 1	–	–	–
ConvT + Tanh	**Ins** + lab_{in} × 1 × 1	256	2,1,0	True
ConvT + BN + LReLU	Lab_2 + 256 × 2 × 2	512	2,2,0	False
Upsample@2	Lab_4+512 × 4 × 4	–	–	–
Conv + BN + LReLU	522 × 8 × 8	256	3,1,1	False
Upsample@2	Lab_8+256 × 8 × 8	–	–	–
Conv + BN + LReLU	266 × 16 × 16	256	5,1,2	False
Conv + BN + LReLU	256 × 16 × 16	128	3,1,1	False
Upsample@2	128 × 16 × 16	–	–	–
Conv + BN + LReLU	128 × 32 × 32	128	5,1,2	False
Conv + BN + LReLU	128 × 32 × 32	64	3,1,1	False
Upsample@2	64 × 32 × 32	–	–	–
Conv + BN + LReLU	64 × 64 × 64	64	3,1,1	False
Conv + Tanh	64 × 64 × 64	**Out**	3,1,1	True
Output	**Out** × 64 × 64	–	–	–

Table 8. Generator for MNIST.

Layer	Input size (C × H × W)	Filters units (width)	K, S, P	bias
label scaler				
Lab_{in}	**Label size** × 1 × 1	–	–	–
Lab_2 ConvT + Tanh	**Label Size** × 1 × 1	**Label Size**	2,1,0	True
Lab_4 ConvT + LReLU	**Label Size** × 2 × 2	**Label Size**	2,2,0	True
Lab_8 ConvT + LReLU	**Label Size** × 4 × 4	**Label Size**	2,2,0	True
Main network				
Input	**Ins** + lab_{in} × 1 × 1	–	–	–
ConvT + Tanh	**Ins** + lab_{in} × 1 × 1	256	2,1,0	True
ConvT + BN + LReLU	Lab_2 + 256 × 2 × 2	512	2,2,0	False
Upsample@2	Lab_4+512 × 4 × 4	–	–	–
Conv + BN + LReLU	522 × 8 × 8	256	3,1,1	False
Upsample@2	Lab_8+256 × 8 × 8	–	–	–
Conv + BN + LReLU	266 × 16 × 16	128	3,1,1	False
Upsample@2	128 × 16 × 16	–	–	–
Conv + BN + LReLU	128 × 32 × 32	64	3,1,1	False
Conv + Tanh	64 × 64 × 64	**Out**	3,1,1	True
Output	**Out** × 32 × 32	–	–	–

Fig. 19. Full feature traversal of a reconstructed image. Columns from left to right: -1, -0.75, -0.5, -0.25, 0, 0.25, 0.5, 0.75, 1. Rows represent the input channels/encoder output channels. Without GDNs.

Fig. 20. Full feature traversal of a generated image. Columns from left to right: -1, -0.75, -0.5, -0.25, 0, 0.25, 0.5, 0.75, 1. Rows represent the input channels/encoder output channels. With GDNs.

Fig. 21. Full attribute traversal of a generated image. Columns from left to right: −1, −0.75, −0.5, −0.25, 0, 0.25, 0.5, 0.75, 1. Rows from top to bottom: Bangs, Black_Hair, Eyeglasses, Heavy_Makeup, Male, Receding_Hairline, Smiling, Straight_Hair, Wavy_Hair, Young. With GDNs.

References

1. Arjovsky, M., Chintala, S., Bottou, L.: Wasserstein generative adversarial networks. In: International Conference on Machine Learning, pp. 214–223. PMLR (2017). https://arxiv.org/pdf/1701.07875.pdf
2. Bank, D., Koenigstein, N., Giryes, R.: Autoencoders. arXiv preprint arXiv:2003.05991 (2020). https://arxiv.org/pdf/2003.05991.pdf
3. Borji, A.: Pros and cons of GAN evaluation measures. Comput. Vis. Image Understanding **179**, 41–65 (2019). https://arxiv.org/pdf/1802.03446.pdf
4. Chadha, G.S., Panara, U., Schwung, A., Ding, S.X.: Generalized dilation convolutional neural networks for remaining useful lifetime estimation. Neurocomputing **452**, 182–199 (2021). https://doi.org/10.1016/j.neucom.2021.04.109. https://www.sciencedirect.com/science/article/pii/S0925231221006676
5. Chadha, G. S., Reimann, J.N., Schwung, A.: Generalized dilation structures in convolutional neural networks. In: Proceedings of the 10th International Conference on Pattern Recognition Applications and Methods, pp. 79–88. ICPRAM (2021). https://doi.org/10.5220/0010302800790088
6. Chen, X., Duan, Y., Houthooft, R., Schulman, J., Sutskever, I., Abbeel, P.: InfoGAN: interpretable representation learning by information maximizing generative adversarial nets. In: Proceedings of the 30th International Conference on Neural Information Processing Systems, pp. 2180–2188 (2016). https://arxiv.org/pdf/1606.03657v1.pdf
7. Choe, J., Park, S., Kim, K., Hyun Park, J., Kim, D., Shim, H.: Face generation for low-shot learning using generative adversarial networks. In: Proceedings of the IEEE International Conference on Computer Vision Workshops, pp. 1940–1948 (2017). https://openaccess.thecvf.com/content_ICCV_2017_workshops/papers/w27/Choe_Face_Generation_for_ICCV_2017_paper.pdf
8. Dai, J., et al.: Deformable convolutional networks. In: Proceedings of the IEEE International Conference on Computer Vision, pp. 764–773 (2017)
9. Donahue, J., Krähenbühl, P., Darrell, T.: Adversarial feature learning. arXiv preprint arXiv:1605.09782 (2016). https://arxiv.org/pdf/1605.09782.pdf
10. Dumoulin, V., et al.: Adversarially learned inference. arXiv preprint arXiv:1606.00704 (2016). https://arxiv.org/pdf/1606.00704.pdf
11. Gonzalez-Garcia, A., Van De Weijer, J., Bengio, Y.: Image-to-image translation for cross-domain disentanglement. arXiv preprint arXiv:1805.09730 (2018). https://arxiv.org/abs/1805.09730.pdf
12. Goodfellow, I.: Nips 2016 tutorial: generative adversarial networks. arXiv preprint arXiv:1701.00160 (2016). https://arxiv.org/pdf/1701.00160.pdf
13. Goodfellow, I., et al.: Generative adversarial nets. In: Advances in Neural Information Processing Systems 27 (2014). https://papers.nips.cc/paper/2014/file/5ca3e9b122f61f8f06494c97b1afccf3-Paper.pdf
14. Guo, X., Liu, X., Zhu, E., Yin, J.: Deep clustering with convolutional autoencoders. In: Liu, D., Xie, S., Li, Y., Zhao, D., El-Alfy, ES. (eds.) Neural Information Processing. ICONIP 2017. LNCS, vol. 10635, pp. 373–382. Springer, Cham (2017). https://doi.org/10.1007/978-3-319-70096-0_39. https://xifengguo.github.io/papers/ICONIP17-DCEC.pdf
15. He, Y., Keuper, M., Schiele, B., Fritz, M.: Learning dilation factors for semantic segmentation of street scenes. In: German Conference on Pattern Recognition, pp. 41–51 (2017)
16. Higgins, I., et al.: Towards a definition of disentangled representations. arXiv preprint arXiv:1812.02230 (2018). https://arxiv.org/pdf/1812.02230.pdf
17. Higgins, I., et al.: beta-VAE: Learning basic visual concepts with a constrained variational framework (2016). https://openreview.net/pdf?id=Sy2fzU9gl

18. Isola, P., Zhu, J.Y., Zhou, T., Efros, A.A.: Image-to-image translation with conditional adversarial networks. In: Proceedings of the IEEE Conference on Computer Vision and Pattern Recognition, pp. 1125–1134 (2017). https://openaccess.thecvf.com/content_cvpr_2017/papers/Isola_Image-To-Image_Translation_With_CVPR_2017_paper.pdf

19. Karras, T., Aila, T., Laine, S., Lehtinen, J.: Progressive growing of GANs for improved quality, stability, and variation. arXiv preprint arXiv:1710.10196 (2017). https://arxiv.org/pdf/1710.10196.pdf

20. Karras, T., Laine, S., Aila, T.: A style-based generator architecture for generative adversarial networks. In: Proceedings of the IEEE/CVF Conference on Computer Vision and Pattern Recognition, pp. 4401–4410 (2019). https://openaccess.thecvf.com/content_CVPR_2019/papers/Karras_A_Style-Based_Generator_Architecture_for_Generative_Adversarial_Networks_CVPR_2019_paper.pdf

21. Kingma, D.P., Welling, M.: Auto-encoding variational bayes. arXiv preprint arXiv:1312.6114 (2013). https://arxiv.org/pdf/1312.6114.pdf

22. Larsen, A.B.L., Sønderby, S.K., Larochelle, H., Winther, O.: Autoencoding beyond pixels using a learned similarity metric. In: International Conference on Machine Learning, pp. 1558–1566. PMLR (2016). http://proceedings.mlr.press/v48/larsen16.pdf

23. Lazarou, C.: Autoencoding generative adversarial networks. arXiv preprint arXiv:2004.05472 (2020). https://arxiv.org/pdf/2004.05472.pdf

24. LeCun, Y., Bottou, L., Bengio, Y., Haffner, P.: Gradient-based learning applied to document recognition. Proceed. IEEE **86**(11), 2278–2324 (1998). https://ieeexplore.ieee.org/stamp/stamp.jsp?tp=&arnumber=726791

25. Ledig, C., et al.: Photo-realistic single image super-resolution using a generative adversarial network. In: Proceedings of the IEEE conference on computer vision and pattern recognition, pp. 4681–4690 (2017). https://openaccess.thecvf.com/content_cvpr_2017/papers/Ledig_Photo-Realistic_Single_Image_CVPR_2017_paper.pdf

26. Liu, M.Y., Breuel, T., Kautz, J.: Unsupervised image-to-image translation networks. In: Advances in Neural Information Processing Systems, pp. 700–708 (2017). https://proceedings.neurips.cc/paper/2017/file/dc6a6489640ca02b0d42dabeb8e46bb7-Paper.pdf

27. Liu, Z., Luo, P., Wang, X., Tang, X.: Deep learning face attributes in the wild. In: Proceedings of International Conference on Computer Vision (ICCV) (2015). https://arxiv.org/pdf/1411.7766.pdf

28. Mahapatra, D., Bozorgtabar, B., Garnavi, R.: Image super-resolution using progressive generative adversarial networks for medical image analysis. Computer. Med. Imaging Graph. **71**, 30–39 (2019). https://www.sciencedirect.com/science/article/abs/pii/S0895611118305871

29. Makhzani, A., Shlens, J., Jaitly, N., Goodfellow, I., Frey, B.: Adversarial autoencoders. arXiv preprint arXiv:1511.05644 (2015). https://arxiv.org/pdf/1511.05644.pdf

30. Mao, X., Li, Q., Xie, H., Lau, R.Y., Wang, Z., Paul Smolley, S.: Least squares generative adversarial networks. In: Proceedings of the IEEE International Conference on Computer Vision, pp. 2794–2802 (2017). https://arxiv.org/pdf/1611.04076.pdf

31. Matthey, L., Higgins, I., Hassabis, D., Lerchner, A.: dsprites: disentanglement testing sprites dataset. https://github.com/deepmind/dsprites-dataset/ (2017)

32. Minhas, M.S., Zelek, J.: Semi-supervised anomaly detection using autoencoders. arXiv preprint arXiv:2001.03674 (2020). https://arxiv.org/pdf/2001.03674.pdf

33. Mirza, M., Osindero, S.: Conditional generative adversarial nets. arXiv preprint arXiv:1411.1784 (2014). https://arxiv.org/pdf/1411.1784v1.pdf

34. Odena, A., Dumoulin, V., Olah, C.: Deconvolution and checkerboard artifacts. Distill (2016). https://doi.org/10.23915/distill.00003. http://distill.pub/2016/deconv-checkerboard

35. Odena, A., Olah, C., Shlens, J.: Conditional image synthesis with auxiliary classifier gans. In: International Conference on Machine Learning, pp. 2642–2651. PMLR (2017). https://arxiv.org/pdf/1610.09585.pdf

36. Perarnau, G., Van De Weijer, J., Raducanu, B., Álvarez, J.M.: Invertible conditional GANs for image editing. arXiv preprint arXiv:1611.06355 (2016). https://arxiv.org/pdf/1611.06355.pdf
37. Radford, A., Metz, L., Chintala, S.: Unsupervised representation learning with deep convolutional generative adversarial networks. arXiv preprint arXiv:1511.06434 (2015). https://arxiv.org/pdf/1511.06434v2.pdf
38. Rosca, M., Lakshminarayanan, B., Warde-Farley, D., Mohamed, S.: Variational approaches for auto-encoding generative adversarial networks. arXiv preprint arXiv:1706.04987 (2017). https://arxiv.org/pdf/1706.04987.pdf
39. Rumelhart, D.E., Hinton, G.E., Williams, R.J.: Learning internal representations by error propagation. Tech. rep., California Univ San Diego La Jolla Inst for Cognitive Science (1985). https://apps.dtic.mil/sti/pdfs/ADA164453.pdf
40. Ulyanov, D., Vedaldi, A., Lempitsky, V.: It takes (only) two: adversarial generator-encoder networks. In: Thirty-Second AAAI Conference on Artificial Intelligence (2018). http://sites.skoltech.ru/app/data/uploads/sites/25/2017/04/AGE.pdf
41. Jeon, Y., Kim, J.: Active convolution: learning the shape of convolution for image classification. In: 2017 IEEE Conference on Computer Vision and Pattern Recognition (CVPR), pp. 1846–1854 (2017)
42. Zhang, Z., Zhang, R., Li, Z., Bengio, Y., Paull, L.: Perceptual generative autoencoders. In: International Conference on Machine Learning, pp. 11298–11306. PMLR (2020). http://proceedings.mlr.press/v119/zhang20ab/zhang20ab.pdf
43. Zhu, J., Shen, Y., Zhao, D., Zhou, B.: In-domain GAN inversion for real image editing. In: Vedaldi, A., Bischof, H., Brox, T., Frahm, J.-M. (eds.) ECCV 2020. LNCS, vol. 12362, pp. 592–608. Springer, Cham (2020). https://doi.org/10.1007/978-3-030-58520-4_35
44. Zhu, J.Y., Park, T., Isola, P., Efros, A.A.: Unpaired image-to-image translation using cycle-consistent adversarial networks (2020)

Retinotopic Image Encoding by Samples of Counts

Viacheslav Antsiperov$^{(\boxtimes)}$ (ID) and Vladislav Kershner

Kotelnikov Institute of Radioengineering and Electronics of RAS, Mokhovaya 11-7,
Moscow 125009, Russian Federation
antciperov@cplire.ru

Abstract. The article is devoted to the synthesis of image encoding methods based on the images data themselves. The proposed approach is based on a previously developed special representation of images by samples of counts (sampling representations). Since the sampling representations are essentially random constructions, the synthesis of encoding methods is carried out strictly within the framework of the generative paradigm. In essence, the approach proposed treats the image coding within generative model as a special case of the classical statistical problem of probability distribution density estimation. In the paper we restrict ourselves to the class of parametric estimation procedures, which imply some parametric family of probability distributions. Namely, we propose to use the model of a parametric mixture of simple distribution components. Accordingly, a set of component weights estimates calculated from a sampling representation, considering as input data, is interpreted as an encoded image – output data. In this context, optimal coding is synthesized with the maximum likelihood method. For the algorithmic implementation of the coding procedure the mixture model is equipped with the structure of receptive fields, that is a well-known organizing principal for receptors in the human eye retina. On this basis, we synthesize a relatively simple recurrent coding algorithm, which turned out to be close to the popular in machine learning EM algorithm. The paper presents interpretation of several features of the algorithm from the point of view of well-known facts about the image processing in the periphery of the visual system, discusses options for the algorithm implementation, and presents the results of numerical simulation of its operation.

Keywords: Perceptual image coding · Generative models · Machine learning

1 Introduction

The principles of transformation, registration, and processing of the light flux in the human/higher vertebrate visual system have been used since the invention of the first photo cameras. In the process of development from the simplest camera obscura to modern digital cameras, the number of borrowed principles has only increased. We could mention here the use of a camera diaphragm as an analogue of the pupil, the optical system of the camera as a crystalline lens, and a registering light element in the

© Springer Nature Switzerland AG 2023
M. De Marsico et al. (Eds.): ICPRAM 2021/2022, LNCS 13822, pp. 52–75, 2023.
https://doi.org/10.1007/978-3-031-24538-1_3

form of a plate with a photographic emulsion, or in the form of a CCD/CMOS matrix as an analogue of the retina with a huge number of photoreceptors, see Fig. 1. The fact that the light-sensitive surface of the eye (retina) has the form of hemisphere and not a plane, is not important for image formation.

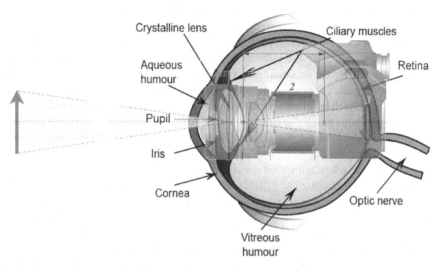

Fig. 1. The correlations of the light transformation/registration methods in the human eye and in the photo camera – in natural and in artificial imaging systems.

It should be noted that the most profound study and implementation of the visual perception mechanisms into artificial imaging systems concerns some aspects of the light registration. It is worth recalling such discoveries as the daguerreotyping – a mercury vapor imaging technology developed in 1837 and its heirs, silver-coated plates, which by the 20th century had been improved into celluloid photographic films with a gelatin-silver emulsion that marked the age of analogue photography. By the beginning of the 21st century, radiation registration technology had risen to a new level, ushering in the era of digital photography, based on matrixes of photodiodes. These advances were primarily due to the invention of charge-coupled devices (CCDs) in 1969 and the subsequent invention of photosensitive matrices based on complementary metal-oxide-conductor (CMOS) structures. In 1993, the first active pixel CMOS sensor was developed [1]. The transition from analogue to digital photography made it possible to significantly improve almost all parameters of imaging systems, among which there are: the increase in spatial resolution, reduction in power consumption, shortening of exposure time, etc.

With the transition to digital imaging devices, it became possible to control the size of individual recording elements – jots, artificial analogues of retinal photoreceptors. To date, the release of a 41 MB matrix with a pixel pitch of 2.2 microns, operating at a frequency of 30 frames per second, has been announced [2]. Considering that the human eye contains ~100 million receptors each of them 2.5–5 microns in size, it becomes clear that modern artificial imaging systems if not already reaching the characteristics of the retina, have come close to it. Note that with such a detector sizes (~1 μm), the radiation registration mechanisms acquire a pronounced quantum character, which leads to the operation of video-matrices in the so-called single photon counting mode (quanta image sensors - QIS). This opens principally new possibilities and raises new challenges in the field of imaging [3].

However, the most important advantage of digital images, which provided them with a quick and unconditional victory over analogues, is their ability to convert the recorded light radiation directly into an output electric current. This photocurrent, without any intermediate accumulation, storage, fixing agents, can be directly transferred to the electrical circuits following the video matrix, which, in turn, includes microprocessors for performing numerous processing/imaging tasks. The computing power of modern microprocessors makes it possible to solve not only a standard set of video-signal pre-processing tasks, such as glare compensation, dark current correction, white balance, etc., but also performs much more complex operations like those that occur in the central visual system of the human brain. The online use of microprocessors in modern digital systems opens almost unlimited possibilities for the formation, processing and analysis of recorded images using numerous data processing methods.

Among the most popular and effective methods of data processing, in the first place, of course, machine learning (ML) should be noted. To date, ML methods, especially those associated with neural networks, have achieved such impressive results, that their widespread implementation is regarded as an artificial intelligence (AI) revolution [4]. The most impressive applications of ML in the field of image processing are the following: handwriting recognition, textual images description, recognition of faces/objects in images, etc. At the same time, paradoxically, the reasons for these results are not entirely clear – often achieved by experimental verification, they still do not have a convincing theoretical explanation. The most you can say with confidence is which of ML methods are most successful in practice.

Since the main feature of machine learning is the focus on data, the characteristics of various methods are determined, firstly, by the specification of the training data, and, secondly, by the choice of a decision (markup)/data model. The specification of training data means the presence or absence of explicit markup in the data (i.e. decision labels, annotations associated with the data). This leads to a division of ML methods into supervised and unsupervised learning. Regarding the used (statistical) model that relates data to the markup, the machine learning methods are divided into discriminant and generative methods, depending on whether the markup model is based on a conditional probability of decision with respect to the data, or on their joint distribution [5]. The history of machine learning development knows examples of all combinations of these dichotomies – supervised/unsupervised learning and discriminant/generative modeling. The experience obtained up to now says that approaches based on generative models and unsupervised learning are more successful in terms of overall quality. Such approaches include generative adversarial networks [6], variational autoencoders [7], deep belief networks [8], etc.

Certainly, the reason for the success of unsupervised generative models is also not fully understood. However, it is increasingly being argued that this may be due to more adequate modeling of the mechanisms of natural (human/higher vertebrate) intelligence [9]. Indeed, as the above models [6–8] improved, every new stage of development added new functions/elements to them, modeling, for example, features of the hierarchical architecture of the cerebral cortex, deep reinforcement learning, working memory in recurrent cortical networks, long–term memory etc. (see details in [9]). Since the details of information processes in the cerebral cortex are not completely clear, it is necessary to model the corresponding functions/elements using computer data representation structures. Hence great importance is attached to the choice of adequate representations. Of course, the modeled functions and structures of ML largely determine the representations of the input – intermediate – output data and (linear/non-linear) relations between them. At the same time, there are several examples where a good choice of data representation significantly improved the efficiency of the functions [6–8]. In this regard, a very important question arises – to what extent can the choice of one or another data representation predetermine the synthesis of the expected functions of the method? In other words, to what extent data representation can determine the development of effective ML methods, just as, in the mid-80s, the object–oriented approach fundamentally changed the software development [10]?

This work provides a partial answer to the question posed. We propose a new approach to the specific ML problem – the problem of synthesizing image encoding methods based on the data of the images themselves. The proposed approach is based on a previously developed special representation of images using a samples of counts (sampling representations) [11, 12]. The main arguments in favor of the usefulness of such representations were discussed in detail in [11] and summarized in the next section. Section 2 also presents a more rigorous derivation of the statistical description of sampling representations from the ideal image model. The convenience of sampling representations for machine learning problems, noted in [11] and demonstrated by the example of smoothing noisy images using the Parzen-Rosenblatt window method [11], is demonstrated in this paper for a more complex problem of perceptual image coding. Section 3 is completely devoted to one of the possible its solutions. Since for sampling representations the complete statistical description is given by the product of the probability distribution densities of its individual counts, the proposed approach is based, in essence, on the classical problem of estimating their probability densities (density estimation) [13]. In the paper we restrict ourselves to the class of parametric estimation procedures [13], which implies a parametric family of probability distributions. Namely, we propose to use the model of a parametric mixture of simple distribution components [14]. Accordingly, a set of component weights estimates calculated from a sampling representation (input data) is considered as an encoded image (output data). In this context, optimal coding is synthesized with the maximum likelihood method [15]. For the algorithmic (computer) implementation of the coding procedure, mixture model is introduced in the form of a system of receptive fields (RF), which are covering the entire image.

2   Image Representation by Sample of Counts (Sampling Representation)

The motivation of the image representation proposed below can be obtained from an analysis of two types of light registering systems: natural, represented by the human eye retina, and artificial, represented by photographic plates, video matrices, etc. It was emphasized in introduction, that the parallels between natural and artificial systems are not accidental, since the structure of the registering elements in almost all imaging systems explicitly or implicitly borrowed the structure of the retina as a 2D–set of photoreceptors. An interesting fact is that the development of such elements from gelatin plates to CMOS matrices was accompanied by an increase in their similarity to the retina. To emphasize it and to make subsequent formal definition of sampling representation and associated models not too formal, we present below some facts about natural and artificial light detecting systems.

The input element in the human imaging system is the retina. It includes about 100 million rods and 10 million cones capable of registering individual photons of the radiation in visible spectrum. The density of photoreceptors in retina varies from 100 to 160 thousand receptors per mm^2 (so the distance or pitch between receptors is ~2–5 microns). The reaction time of single photoreceptor is about or greater than 20 ms. But it is worth noting that the signals going to the brain via the optic nerve are not the same as those registered by photoreceptors in outer layer. These signals are formed in a complex system of cells of the middle and inner retina layers, after which they come along the set of ganglion cell axons composing optic nerve to the visual cortex. The number of axons is about a million, which is about two orders of magnitude less than the number of photoreceptors.

The technical implementation of the registering element in modern photo cameras is the CMOS video matrix. As it was announced in the Gigajot company press release (Apr 04, 2022) [2], the following characteristics have been achieved for implemented in CMOS technology Quanta image sensor (QIS) – a room-temperature photon-counting sensor without avalanche multiplication: 41 Megapixel matrix utilizes a 2.2-micron pixel and has a read noise of only 0.35 electrons, photon counting and photon number resolving up to its top speed of 30 frames per second at full resolution. The high resolution and the extremely low read noise provide flexibility for binning and additional post-processing, while maintaining a read noise that is still lower than native lower resolution sensors.

Comparing the artificial and natural registering systems, it is easy to outline their common features. Both have a finite photosensitive 2D–surface, containing a huge number of receptors/photodetectors. All detectors can register single photons of incident radiation. Both systems memorize for a short period of time (frame) the events associated with the photons registration. The noted set of characteristics can be used to formalize the concept of the imaging device, which generalizes not only the above systems, but also several others, including photographic films, plates, etc. (not only in the visible spectrum).

Formally, by an ideal imaging device, we further understand a flat 2D–surface Ω on which identical point detectors, or "jots" in terms of [1], are located close to each other. Point detectors, by definition, have a photosensitive surfaces with a very small area ds. Accordingly, the total number of detectors will be $N = S/ds$, where S is the total area of surface Ω. Assuming that S is fixed and $ds \to 0$, the number N is assumed to be very large: $N \to \infty$. Thus, an ideal imaging device is an almost continuous sensitive surface Ω with coordinates $\vec{x} = (x_1, x_2)$, that define the positions of ideal point detectors, as shown in Fig. 2.

Based on the concept of an ideal imaging device, an ideal image model can be defined. An ideal image is understood as an (ordered) set $X = (\vec{x}_1, \ldots, \vec{x}_n)$ of all n random counts registered by point detectors of an ideal imaging device with coordinates $\vec{x}_1, \ldots, \vec{x}_n$, during the frame time T. An ideal image is, therefore, a fundamentally random object. The random nature of an ideal image is determined not only by random coordinates $\vec{x}_i$ of counts, but also by their random total number n (size of X).

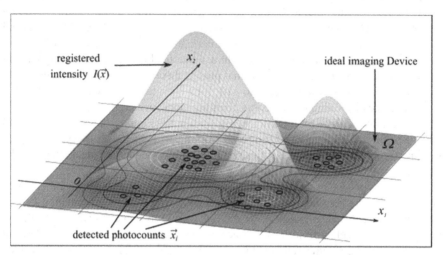

Fig. 2. An ideal imaging device Ω and the result of radiation intensity $I(\vec{x})$ registration in the form of a set of photocounts $X = (\vec{x}_1, \ldots, \vec{x}_n)$.

A complete statistical description of ideal image in the form of a set of all multi-variate distribution densities $\{\rho(\vec{x}_1, \ldots, \vec{x}_n, n | I(\vec{x}))\}$ can be obtained by assuming conditional independence of counts $\vec{x}_i$ (for a given registered intensity $I(\vec{x})$). The detailed statistical deduction of this description is presented in [11]. Alternative deduction, based on the Poisson approximation ($ds \to 0, S = const, N \to \infty$) of the joint probability of a large ensemble of Bernoulli point detectors, registering counts with the probability (success probability) $P(\vec{x} | I(\vec{x})) = \alpha T I(\vec{x})$, can be found, for example, in [16]. Here we present only the result, which establishes that the random set of counts $X = (\vec{x}_1, \ldots, \vec{x}_n)$ coincides with the 2D Poisson point process (PPP) on Ω, which intensity $\bar{n}(\vec{x})$ (average number of counts per unit area) up to a factor αT coincides with the intensity $I(\vec{x})$ of radiation registered ($\bar{n}(\vec{x}) = \alpha T I(\vec{x})$):

$$\rho(\vec{x}_1, \ldots, \vec{x}_n, n | I(\vec{x})) = \rho(\vec{x}_1, \ldots, \vec{x}_n, | n, I(\vec{x})) \times P_n(I(\vec{x}))$$
$$= \prod_{i=1}^{n} \rho(\vec{x}_i | I(\vec{x})) \times P_n, \tag{1}$$
$$P_n = \frac{\bar{n}^n}{n!} \exp(-\bar{n}), \bar{n} = \iint_\Omega \bar{n}(\vec{x}) ds = \alpha T \iint_\Omega I(\vec{x}) ds,$$

where the density of single count $\rho(\vec{x}_i | I(\vec{x}))$ has the form:

$$\rho(\vec{x}_i | I(\vec{x})) = \frac{\bar{n}(\vec{x}_i)}{\bar{n}} = \frac{I(\vec{x}_i)}{\iint_\Omega I(\vec{x}) ds} \tag{2}$$

In the above expressions, the parameter $\alpha = \eta(h\bar{v})^{-1}$ depends on $h\bar{v}$ – the average energy of the detecting photon (h is Planck's constant, $\bar{v}$ is the characteristic radiation frequency) and on dimensionless coefficient η – the quantum efficiency of detector. Questions of substantiation of the given model of registration, based on the modern (quantum) physics, can be found in [17].

The model of an ideal image and its statistical description (1–2) are certainly not something original, they have long been commonly used at low intensities $I(\vec{x})$ of registered radiation, for example, in the fields of fluorescence microscopy, positron emission tomography (PET), single photon emission computed tomography (SPECT), optical and infrared astronomy, etc. [18]. However, under normal conditions, with radiation intensities corresponding, for example, to day light, the practical use of an ideal image model turns out to be problematic. The fact is that photon fluxes, for example, on a clear day from the sun are huge – they amount to $\sim 10^{15}$–10^{16} photons per area $S \sim \text{mm}^2$ in 1 s. For ideal devices operating in the photon counting mode, even if they register one count per ~ 10 photons (with quantum efficiency $\eta = 0.1$), the number of counts per second will be $\bar{n} \sim 10^{15}$ (1 Peta-count). Obviously, working with such data flows will require too many resources. Therefore, it is desirable to somehow adapt the ideal image model to real practical conditions.

Some time ago, we proposed the following solution to the above problem [11, 12]. Let us fix from the very beginning some acceptable size of the representation $k \ll \bar{n}$ and, considering the ideal image $X = (\vec{x}_1, \ldots, \vec{x}_n)$ as some statistical population of random counts, select from it, in full accordance with the approach of the classical statistical theory, a random a sample of k counts $X_k = (\vec{x}_{j_1}, \ldots, \vec{x}_{j_k})$. Obviously, such a "subset" still, although with a much smaller size $k \ll \bar{n}$, represents the original (ideal) image X. Let us call this X_k the image representation by a sample of random counts or, in short, a sampling representation. The statistical description of sampling representation easily follows from (1) by integrating $\rho(\vec{x}_1, \ldots, \vec{x}_n, n | I(\vec{x}))$ over counts not selected in X_k and summing over the number $l = n - k = 0, 1, \ldots$ of unselected counts:

$$\rho\left(X_k | I(\vec{x})\right) = \prod_{j=1}^{k} \rho(\vec{x}_j | I(\vec{x})) \times P_{n \geq k},$$
$$P_{n \geq k} = \sum_{l=0}^{\infty} P_{k+l}. \tag{3}$$

where $P_{n \geq k}$ denotes the probability that an ideal image contains more than k counts. As it is known, for the Poisson distribution P_n (1), probability $P_{n \geq k}$ is equal to ratio of incomplete gamma function to complete gamma function [19] $P_{n \geq k} = \gamma(k, \bar{n})/\Gamma(k)$. Using in the case $k \ll \bar{n}$ for $\gamma(k, \bar{n})$ approximation $\sim \left[\Gamma(k) - \bar{n}^{(k-1)} \exp\{-\bar{n}\}\right]$ and for $\Gamma(k)$ the Stirling's approximation [19], we can find the following estimate:

$$P_{n \geq k} \approx 1 - \frac{1}{\sqrt{2\pi(k-1)}} \frac{\bar{n}^{k-1} \exp\{-\bar{n}\}}{(k-1)^{k-1} \exp\{-(k-1)\}} > 1 - \frac{1}{\sqrt{2\pi(k-1)}} \tag{4}$$

which follows from monotonic decrease of the function $\xi^{k-1} \exp\{-\xi\}$ for $\xi \geq k - 1$.

It follows from the estimate (4) that the probability $P_{n \geq k}$ differs from unity by a small value not exceeding $1/\sqrt{2\pi(k-1)}$, and therefore, under the assumption implied $1 \ll k \ll \bar{m}$, it can be set equal to one. This allows us to rewrite the statistical description of the sampling representation X_k (3) in the following independent of $\bar{n}$ form:

$$\rho(X_k | I(\vec{x})) = \prod_{j=1}^{k} \rho(\vec{x}_j | I(\vec{x}))$$
$$\rho(\vec{x}_j | I(\vec{x})) = \frac{I(\vec{x}_j)}{\iint_\Omega I(\vec{x}) ds} \tag{5}$$

where the indexing of counts $\vec{x}_j$ is introduced internally within the framework of the sampling representation X_k by means of the $j = 1, \ldots, k$.

Statistical description (5) for sampling representation $X_k = (\vec{x}_1, \ldots, \vec{x}_k)$ has several remarkable properties in terms of its subsequent processing/utilization. First, (5) fixes the conditional independence and the identical distribution (iid property) of all k counts of the representation. Secondly, the distribution density of each of the counts $\rho(\vec{x}_j | I(\vec{x}))$ is related to the intensity of the detected radiation $I(\vec{x})$ in the simplest way – it is equal to normalized intensity. And, thirdly, description (5) is, in a certain sense, universal – it does not depend on the details of the registration mechanism (jots of CMOS matrix, receptors of retinal, etc.), namely on the quantum efficiency η, or on the spectrum of the registered radiation, or on the frame time T. These properties characterize sampling representations as an extremely convenient form of input data representation for many well-developed and proven machine learning methods [20].

Moreover, since $\rho(\vec{x}_j | I(\vec{x}))$, and hence $\rho(X_k | I(\vec{x}))$ (5) do not depend on the absolute values of the intensity, but are determined by its normalized version, the statistical description of sampling representation also does not depend on the chosen units of $I(\vec{x})$. So, if the intensity of the registered radiation is given by pixels $\{n_i\}$ of some digital image, the description (5) will not depend on the quantization resolution parameter $Q = \Delta I$, but only on the pixel bit depth υ. In this regard, we note that the procedure for generating a sampling representation for digital images can essentially be reduced to the normalization $\pi_i = n_i / \sum n_i$ of the pixel values $n_i \sim I_i / Q$ of the image and the subsequent sampling of k counts from resulting probability distribution $\rho(\vec{x}_j | I(\vec{x})) \approx \pi_i$, where the count $\vec{x}_j$ belongs to the surface $d\Omega$ of the pixel n_i. It should be noted that in the field of machine learning there is a whole arsenal of methods for organizing sampling procedures, united by the common name Monte Carlo methods [21].

To illustrate the Monte Carlo method for generating sample representations of digital images and to get a general idea of the quality of such representations, let us consider representations of the standard image "GRAY_R02_0600x0600_093.png" from the TESTIMAGES archive. [22] (see Fig. 3 A).

In this example, a rejection sampling algorithm was used [21]. The algorithm is executed iteratively, at each iteration generating a random vector $\vec{x} \in \Omega$, having *floating* point coordinates x_1, x_2. The coordinates $x_1 = l_1 * z_1$ and $x_2 = l_2 * z_2$ are obtained by multiplying the corresponding image sizes l_1, l_2 by random numbers z_1, z_2, uniformly distributed over $[0,1)$, which are generated by a standard random number generator. For given $\vec{x}$, the value n_i of the pixel in the row *round* (x_1) and in the column *round* (x_2) of the image bitmap is determined, where *round* (x) is the rounding operation. After that an auxiliary random value $th = 2^\upsilon * z_3$ is produced, where υ is the pixel bit depth of the image and z_3 is standard, uniformly distributed over $[0,1)$ random number. If the pixel value n_i is less than threshold th, the count $\vec{x}$ is accepted to the sample X_k, if not, it is rejected. When the number of counts in X_k becomes equal to given k, the algorithm stops, and the generated counts sample is considered as sampling representation of the image. Figure 3 fragments B-D show examples of sampling representations of 500,000, 1,000,000 and 5,000,000 counts, respectively. Note, that since the B-D images in Fig. 3, showing sampling representations, are physically also bitmap images having the same dimensions as the original one, each their pixel is equal to the number of counts that fell into it, i.e. in a certain sense, B-D can be considered as the results of some representations "digitization".

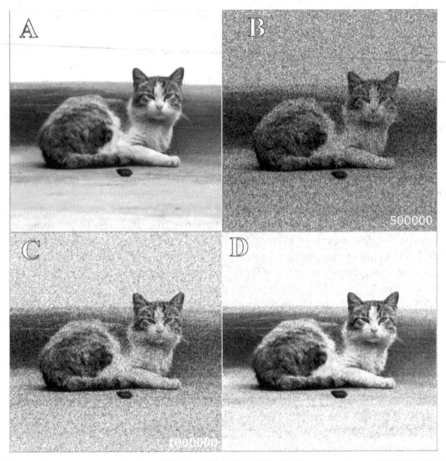

Fig. 3. Representations of the "GRAY_R02_0600x0600_093" image from TESTIMAGES archive [22] by samples of random counts: A – original image in PNG format, B, C, D – sampling representations of sizes, respectively 500.000, 1.000.000 and 5.000.000 counts.

3   Retinotopic Encoding of the Sampling Representation as the Image Compression Method (Perceptual Coding)

Let us use the proposed sampling representations to solve one of the most important problems in the field of image processing – the problem of coding methods synthesis. The synthesis of image coding, which in a broad sense implies their compression, is usually carried out on one of two platforms. The first of them considers the compression as the elimination of statistical (or deterministic) redundancy in the original image, the other – as optimal vector quantization of images, matching with the properties of the human visual system [23]. The goal of both approaches is to achieve the most compact digital representation of images, which would provide maximum economy of resources during their transmission or storage (with limited channel bandwidth or with limited storage media) but would guarantee a given level of possible loss of information (a given quality level).

In this paper, according to the declared focus on modeling the mechanisms of visual system, the second approach is taken as the basis – the synthesis of optimal quantization methods, or, as it is commonly referred to in the field of video/image processing, the synthesis of perceptual coding [24]. The main difference of perceptual coding from encoders reducing redundancy in images is the choice of the similarity metric for the encoded and decoded images [25]. As a rule, traditional coding methods use a metric based on the mean square error (MSE) – the L_2 norm of the difference between the intensities of both images. The MSE metric is easy to calculate, but it is known that it is largely indifferent to image features significant for the human perception. Therefore, although perception-centric coding uses more complex metrics, which are more computationally intensive, the proximity of perception-centric encoders to human intelligence implies a more attractive path for the development of image coding methods [25]. This is confirmed by the noted impressive results in the development of variational autoencoders [7].

To motivate the perceptual metric proposed below for encoding images, we will briefly describe the known facts of visual perception, which mainly concern the periphery of the visual system (the retina) [26]. The retina of the eye is organized in layers. In the outer layer, there are photoreceptor cells – rods and cones, which, under the action of incident light, activate the bipolar cells lying in the middle layer. Bipolar cells transmit receptor signals to ganglion cells located in the inner layer. Ganglion cells, the only ones of all retinal neurons capable of generating an action potential, form nerve impulses in response to received signals and transmit them along the axons collected in the optic nerve to the central visual system (to the brain). It is important to note here that in most cases the nerve impulses sent to the brain are not the same data directly recorded by photoreceptors. They are formed by the retina with the help of numerous intermediate neurons of the middle and inner layers. Among them, in addition to bipolar cells, which carry out vertical connections from the outer to the inner layer, an important role is played by horizontal and amacrine cells, which carry out horizontal connections in the layers. As a result, each ganglion cell can receive signals from dozens and sometimes thousands of receptor cells. A direct consequence of this is an increase in the photosensitivity of ganglion cells to variations in the intensity of the incident radiation. The reverse side of such aggregation is, obviously, a decrease in the spatial resolution of the image neural representation transmitted to the brain. In this regard, it seems that the main purpose of the retina is the optimal resolution of the main problem of communication theory – minimization of data distortions with a limited channel bandwidth [26]. For these purposes, certain mechanisms of primary processing (coding) of photocounts recorded by receptors are implemented in the retina, primarily compression of the stream of recorded events by two orders of magnitude, i.e. ~100 times. Considered in this context, the small areas of the retina, which include photoreceptors associated with a different ganglion cells and provide preprocessing of the corresponding counts, are defined as the receptive fields [26].

The concept of a receptive field (RF) was introduced in 1953 by S. Kuffler [27] and is currently one of the central concepts in neurobiology both in the case of the visual system and in relation to the entire sensory system. As for the retina itself, the details of its RF structure have been elucidated experimentally in the last decade using a special technique that makes it possible to trace the distribution of activation signals from an individual receptor to an array of ganglion cells associated with it [28]. Briefly, the results obtained can be summarized as follows (see Fig. 4). Individual receptors may belong to several RFs, which may differ both in size and function. In turn, the types (functions) of RFs are determined by the types of ganglion cells associated with them. Although today the number of different ganglion cells fits into ~20 types, most cells (~75%) belong to only two of them – midget and parasol types, each with subtypes of ON- and OFF-cells [26]. The cells of the midget type are distinguished by the smaller size of the RF and are responsible for assessing the spatial distribution of the radiation intensity on the RF (see Fig. 4). Parasol cells are characterized by largeRF sizes and are responsible for the change (movement, temporal dynamics) of intensity in the field. Cells of both types are subdivided into ON- and OFF-cells according to the nature of their activation in response to illumination/darkening of their centro-antagonistic RFs [27]. Namely, ON-cells are activated upon stimulation of receptors in the center of RF and inhibited upon stimulation of receptors in the concentric surround, while OFF cells, on the contrary, are activated upon stimulation of receptors in the surround and inhibited upon stimulation of receptors in the center of RF. In the most common models of RF, equal power of photoreceptor stimulation creates a Gaussian spatial activity profile in the narrow center and a slightly broader concentric Gaussian inhibition profile in the antagonistic surround. This model successfully reproduces ganglion cell responses to light spots, grids, and rough chessboards. However, it should be noted that several modern studies have found significant deviations from Gaussian models [28].

As for the RF sets of each of the ganglion cell types, they are all organized in a semi-regular mosaic, uniformly covering the surface of the retina. Moreover, local spatial heterogeneities in the neighboring RFs of the same type complement each other like pieces of a puzzle [30], indicating a finely tuned pattern of their arrangement, covering the entire visual field without gaps. Note that a similar coordination in the location of the RF of different types of ganglion is also observed. Functionally, this coordination appears to make information processing more efficient, allowing for more uniform sampling of the visual world and decorrelation of ganglion cell signals. Figure 5 shows in schematic way the results of experimental measurements of RF locations and shapes in a large populations of ganglion cells [30, 31].

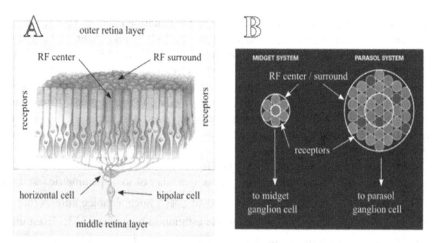

Fig. 4. Schematic representation of the receptive fields. A) Bipolar cell in middle retina layer receives direct synaptic input from a cluster of photoreceptors, constituting the RF center. In addition, it receives indirect input from surrounding photoreceptors via horizontal cells. B) Schematic layout of the connections of cones forming the receptive fields of different types (midget and parasol) ganglion cells. Adapted from [26] and [29].

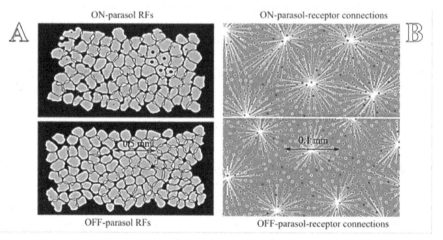

Fig. 5. Locations and shapes of RFs in large populations of ON- and OFF-parasol cells on the retina surface. A) The RFs of ON- and OFF-cells as a regularly spaced mosaic, represented by a collection of contour lines. B) The RFs of ON- and OFF-cell as a connections with the receptors identified in a single recording of the cell sampling. Adapted from [30] and [31].

Let us use the facts presented above to synthesize perceptual methods for encoding recorded images. Since, within the framework of this paper, k-samples of random counts – sampling representation $X_k = (\overrightarrow{x}_1, \ldots, \overrightarrow{x}_k)$ are considered as input data, generated by the probability density distribution (5), it is quite reasonable to understand the input image as the distribution density $\rho(\overrightarrow{x}_j | I(\overrightarrow{x}))$, $\overrightarrow{x} \in \Omega$ (which, up to a norm factor coincides with the representation of the image by the radiation intensity $I(\vec{x})$, see (5)). This automatically brings us to generative data models [32]. Unlike traditional encoders, which are most naturally interpreted as diminishing redundancy operators, encoders in the generative paradigm consider the code as variables associated with a probability distribution, and the encoding operation as a statistical inference procedure (finding probability distribution variables by a given sampling data X_k). In this regard, generative models learn to restore from training sample observations $X_k = (\vec{x}_1, \ldots, \vec{x}_k)$ the probability distribution $\rho(\vec{x}_j)$, that generate them, rather than map the input data to the output with compression.

Considering the above clarification, let us form a model of the desired estimation of probability distribution density $\rho(\vec{x}_j | I(\vec{x}))$ as a member of some parametric family of distributions $\mathcal{G} = \left\{\rho(\vec{x}; \vec{\theta})\right\}$, $\vec{x} \in \Omega$, $\vec{\theta} \in \Theta \subset \mathbb{R}^p$. Such a choice implies that we restrict the following discussion to parametric estimation procedures [13]. Even more specifically: we would restrict ourselves in the search for optimal parameters $\vec{\theta}^{ML}$ by the maximum likelihood method of R. Fisher [15]:

$$
\begin{aligned}
\vec{\theta}^{ML} &= \underset{\vec{\theta} \in \Theta}{arg\,max}\,\ln\rho\left(X_k; \vec{\theta}\right) \\
\rho\left(X_k; \vec{\theta}\right) &= \prod_{j=1}^{k}\rho(\vec{x}_j; \vec{\theta}),
\end{aligned}
\tag{6}
$$

where, in full accordance with (5), the joint distribution $\rho(X_k; \vec{\theta})$ of sampling representation X_k is decomposed into the product of distributions of iid counts $\rho(\vec{x}_j; \vec{\theta})$.

In the light of the receptive fields concept discussed above, let us refine the parametric family $\mathcal{G}$ in the form of a mixture of components corresponding to RFs in the sense described below:

$$
\rho\left(\vec{x}; \vec{\theta}\right) = \sum_{i=1}^{K}\pi_i\rho_i\left(\vec{x}; \vec{\theta}\right),
\tag{7}
$$

where K denotes the number of mixture components, $\{\pi_i\}$, $\pi_{ii} \geq 0$, $i = 1, \ldots, K$ are normalized weights that satisfy the condition $\sum_{i=1}^{K}\pi_i = 1$ and $\left\{\rho_i(\vec{x}; \vec{\theta})\right\}$, $\rho_i(\vec{x}; \vec{\theta}) \geq 0$, $i = 1, \ldots, K$ are the probability distribution densities of count $\vec{x}$ for all components.

To relate the components in (7) to receptive fields, we make the following assumptions. Let us associate each component with some RF so that the corresponding area of that field in the retina would be the carrier $\Delta_i = \{\vec{x} | \rho_i(\vec{x}; \vec{\theta}) \neq 0\}$ of the component probability distribution. It assumes the finite carriers Δ_i of all distribution densities and the covering of surface Ω by the set $\{\Delta_i\}$, $i = 1, \ldots, K$. In addition, we simplify the model (7) by assuming that only two components $\rho_{ic}(\vec{x})$ and $\rho_{is}(\vec{x})$ are associated with any RF. We will interpret them as the centers and surrounds of the corresponding RFs of some selected ganglion cell type (midget, parasol). Moreover, we accept that components $\rho_{ic}(\vec{x})$ and $\rho_{is}(\vec{x})$ are dependent only on the RF index i (from now on, the index i used for numbering the receptive fields) and on no other parameters. Under these assumptions, the model (7) takes the form:

$$\rho\left(\overrightarrow{x};\overrightarrow{\theta}\right) = \sum_{i=1}^{K} w_i \rho_{ic}\left(\overrightarrow{x}\right) + v_i \rho_{is}\left(\overrightarrow{x}\right),$$
$$\sum_{i=1}^{K} w_i + v_i = 1. \tag{8}$$

where $\overrightarrow{\theta}$ hereinafter is a set of weights $\{(w_i, v_i)\}$.

Considering the mixture (8) as a law of total probability, one can interpret the weights w_i and v_i as the probabilities of registering count $\overrightarrow{x}$ in the center or in the surround of the i-th RF, and $\rho_{ic}\left(\overrightarrow{x}\right)$ and $\rho_{is}\left(\overrightarrow{x}\right)$ as conditional distribution densities, provided that $\overrightarrow{x}$ belongs to the center or surround of the given RF. Mixture (8), in turn, can be represented as mixture (7) (but with the index i numbering the receptive fields) denoting $w_i + v_i = \pi_i$ and introducing the conditional probabilities $\pi_{ic} = w_i/\pi_i$ and $\pi_{is} = v_i/\pi_i$ of hitting $\overrightarrow{x}$ to the center or surround of i-th RF, provided that it already belongs to this field. So, introducing unconditional probability density $\rho_i\left(\overrightarrow{x}\right) = \pi_{ic}\rho_{ic}\left(\overrightarrow{x}\right) + \pi_{is}\rho_{is}\left(\overrightarrow{x}\right)$, we will immediately get the mixture (7).

Let us note once again that model (8) relates the $\rho_{ic}\left(\overrightarrow{x}\right)$ and $\rho_{is}\left(\overrightarrow{x}\right)$ to the locations of receptive fields $\{\Delta_i\}$ and to the positions of centers and surrounds in them, but not to the ON-OFF types of these fields. However, it is possible to give an interpretation of (8) in terms of ON- and OFF-RFs, if we use some concepts of multiresolution analysis [33]. Namely, we introduce some fixing conditional probabilities π_{ic}^0 and π_{is}^0, $\pi_{ic}^0 + \pi_{is}^0 = 1$, $i = 1, \ldots, K$ such that corresponding $\rho_i^0 = \pi_{ic}^0\rho_{ic}\left(\overrightarrow{x}\right) + \pi_{is}^0\rho_{is}\left(\overrightarrow{x}\right)$ will be the smoothest density in all mixtures $\pi_{ic}\rho_{ic}\left(\overrightarrow{x}\right) + \pi_{is}\rho_{is}\left(\overrightarrow{x}\right)$ on Δ_i. Then (8) can be rewritten as:

$$\rho\left(\overrightarrow{x};\overrightarrow{\theta}\right) = \sum_{i=1}^{K} \pi_i \rho_i^0\left(\overrightarrow{x}\right) + \delta_i D_i\left(\overrightarrow{x}\right),$$
$$D_i\left(\overrightarrow{x}\right) = \rho_{ic}\left(\overrightarrow{x}\right) - \rho_{is}\left(\overrightarrow{x}\right),$$
$$\delta_i = \pi_i(\pi_{ic} - \pi_{ic}^0) = \pi_i(\pi_{is}^0 - \pi_{is}),$$
$$\pi_i = w_i + v_i. \tag{9}$$

where it is natural to consider the sum of $\pi_i \rho_i^0\left(\overrightarrow{x}\right)$ as the (smooth) approximation of $\rho\left(\overrightarrow{x};\overrightarrow{\theta}\right)$, and the sum of $\delta_i D_i\left(\overrightarrow{x}\right)$ as its details [33]. Unfortunately, the function $D_i\left(\overrightarrow{x}\right)$, unlike $\rho_i^0\left(\overrightarrow{x}\right)$, is not a distribution density (even not positive function), which somewhat violates the generative paradigm. However, representation (9) is often used in image processing, for example, in the form of image processing operator proposed in [34], that is selective for a texture made up of dots or spots.

However, it is possible to correct the noted shortcoming in (9) and come to ON-OFF-interpretation of (8), close to center-surround selective operator [34], if we use the approximation:

$$\rho_i^0\left(\overrightarrow{x}\right) \approx \pi_{ic}^0\rho_{ic}\left(\overrightarrow{x}\right) + \pi_{is}\rho_{is}\left(\overrightarrow{x}\right) \approx \pi_{ic}\rho_{ic}\left(\overrightarrow{x}\right) + \pi_{is}^0\rho_{is}\left(\overrightarrow{x}\right), \tag{10}$$

and rewrite (9) once again in the form:

$$\rho\left(\overrightarrow{x};\overrightarrow{\theta}\right) = \sum_{i=1}^{K} \pi_i \rho_i^0\left(\overrightarrow{x}\right) + r_+(\delta_i)\rho_{ic}\left(\overrightarrow{x}\right) + r_-(\delta_i)\rho_{is}\left(\overrightarrow{x}\right) \tag{11}$$

where $r_+(\delta_i) = \max(\delta_i, 0)$ and $r_-(\delta_i) = \max(-\delta_i, 0)$ are positive rectifying functions, only one of which is non-zero (depending on the sign of δ_i). In accordance with this refinement, the last two terms in (11) can be interpreted as contributions to the i-component from ON- and OFF- corresponding receptive fields.

Returning to the original model (8), we can now write the joint distribution $\rho(X_k; \overrightarrow{\theta})$ of sampling representation X_k and corresponding likelihood function:

$$
\ln\rho\left(X_k; \overrightarrow{\theta}\right) = \ln\prod_{j=1}^{k}\rho(\overrightarrow{x}_j; \overrightarrow{\theta}) = \sum_{j=1}^{k}\ln\rho(\overrightarrow{x}_j; \overrightarrow{\theta})
$$
$$
= \sum_{j=1}^{k}\ln\left(\sum_{i=1}^{K}w_i\rho_{ic}(\overrightarrow{x}_j) + v_i\rho_{is}(\overrightarrow{x}_j)\right)
\tag{12}
$$

The maximum-likelihood parameters $\overrightarrow{\theta}^{ML} = \{(w_i^{ML}, v_i^{ML})\}$, corresponding to likelihood function (12), can be found using the Lagrange multiplier method, considering the only constraint $\sum_{i=1}^{K}w_i + v_i = 1$ (8). The result is a system of equations:

$$
\frac{\partial\ln\rho\left(X_k; \overrightarrow{\theta}\right)}{\partial w_l} = \sum_{j=1}^{k}\frac{\rho_{lc}(\overrightarrow{x}_j)}{\sum_{i=1}^{K}w_i^{ML}\rho_{ic}(\overrightarrow{x}_j)+v_i^{ML}\rho_{is}(\overrightarrow{x}_j)} = \lambda
$$
$$
\frac{\partial\ln\rho\left(X_k; \overrightarrow{\theta}\right)}{\partial v_l} = \sum_{j=1}^{k}\frac{\rho_{ls}(\overrightarrow{x}_j)}{\sum_{i=1}^{K}w_i^{ML}\rho_{ic}(\overrightarrow{x}_j)+v_i^{ML}\rho_{is}(\overrightarrow{x}_j)} = \lambda
\tag{13}
$$

The Lagrange multiplier λ in (13) can be found as follows. We multiply each of the Eqs. (13) by w_l and v_l, respectively, add them up and sum these sums over all l. By virtue of constraint $\sum_{i=1}^{K}w_i + v_i = 1$ there will be λ on the right side, and k on the left, as it is not difficult to see. Therefore, $\lambda = k$, which allows us to rewrite the Eqs. (13) for maximum-likelihood parameters in the following form, which does not contain Lagrange multiplier:

$$
\frac{1}{k}\sum_{j=1}^{k}\frac{\rho_{lc}(\overrightarrow{x}_j)}{\sum_{i=1}^{K}w_i^{ML}\rho_{ic}(\overrightarrow{x}_j)+v_i^{ML}\rho_{is}(\overrightarrow{x}_j)} = 1
$$
$$
\frac{1}{k}\sum_{j=1}^{k}\frac{\rho_{ls}(\overrightarrow{x}_j)}{\sum_{i=1}^{K}w_i^{ML}\rho_{ic}(\overrightarrow{x}_j)+v_i^{ML}\rho_{is}(\overrightarrow{x}_j)} = 1
\tag{14}
$$

In the general case of arbitrary forms and allocations on surface Ω of $\rho_{ic}(\overrightarrow{x})$ and $\rho_{is}(\overrightarrow{x})$, the analytical solution of system (14) is problematic. However, it can be reduced to a convenient for applying numerical recurrent methods form, like the methods for fixed points approximation of equations $\overrightarrow{\theta} = F(\overrightarrow{\theta})$, $\overrightarrow{\theta} = \{(w_i, v_i)\}$ [35], if we multiply Eqs. (14) by w_l and v_l, and rewrite the resulting system in the iterative form:

$$
w_l^{(n+1)} = \frac{1}{k}\sum_{j=1}^{k}\frac{w_l^{(n)}\rho_{lc}(\overrightarrow{x}_j)}{\sum_{i=1}^{K}w_i^{(n)}\rho_{ic}(\overrightarrow{x}_j)+v_i^{(n)}\rho_{is}(\overrightarrow{x}_j)};
$$
$$
v_l^{(n+1)} = \frac{1}{k}\sum_{j=1}^{k}\frac{v_l^{(n)}\rho_{ls}(\overrightarrow{x}_j)}{\sum_{i=1}^{K}w_i^{(n)}\rho_{ic}(\overrightarrow{x}_j)+v_i^{(n)}\rho_{is}(\overrightarrow{x}_j)};
\tag{15}
$$

where $n+1$ denotes the number of current iteration. Note that if we interpret the terms in sums (15) as the estimates of a posteriori probabilities for l-th RF center or surround to contain count $\overrightarrow{x}_j$:

$$
p_{lc|j}^{(n+1)} = \frac{w_l^{(n)}\rho_{lc}(\overrightarrow{x}_j)}{\sum_{i=1}^{K}w_i^{(n)}\rho_{ic}(\overrightarrow{x}_j)+v_i^{(n)}\rho_{is}(\overrightarrow{x}_j)};
$$
$$
p_{ls|j}^{(n+1)} = \frac{v_l^{(n)}\rho_{ls}(\overrightarrow{x}_j)}{\sum_{i=1}^{K}w_i^{(n)}\rho_{ic}(\overrightarrow{x}_j)+v_i^{(n)}\rho_{is}(\overrightarrow{x}_j)};
\tag{16}
$$

then relations (15) will be the M-step of the well-known EM-algorithm (while (16) will be E-step) [36]. It is well known that the EM-algorithm, when the number of components K is relatively small (~10–100) is quite stable and allows calculating maximum-likelihood parameters $\vec{\theta}^{ML}$ in a reasonable time. Unfortunately, with large amounts of data ($k \sim 10^7$ counts) and for a sufficiently large model dimension ($K \sim 10^6$ RFs), the use of the traditional EM algorithm turns out to be ineffective. The high memory requirements $k \times K \sim 10^{12}$ bytes (1 Terabyte) and, accordingly, a large amount of calculations, as well as with a low convergence rate of the (linear) EM algorithm [37] make its traditional computational scheme poorly implemented. In this regard, it seems extremely important that the concept of receptive fields makes it possible to significantly adapt the algorithm (16) to the case of very large input data X_k such as sampling representation of images. An adaptation scheme for iterative calculations of maximum-likelihood parameters $\vec{\theta}^{ML}$, based on a lattice partition of the visual field Ω, was proposed in [12] and consists in the following simplifying assumptions about the locations and shapes of receptive fields.

First, it is assumed that components of mixture (8) are placed at the nodes $\{\vec{\mu}_i\}$, $i = 1, \ldots, K$ of some regular (rectangular or hexagonal) lattice on the surface of Ω. It is assumed that the carriers Δ_i of the adjacent RFs complement each other in such a way, that their covering of the entire visual field has no gaps. This is ensured by the requirement $D > d$, where D is the characteristic size of RF and d is the lattice spacing. Under this condition, obviously, every point $\vec{x} \in \Omega$ belongs to at least one of the carriers. Hence, the set of nodes whose carriers contain $\vec{x}$ is not empty. We denote the nonempty set of indices of these nodes by $\delta_{\vec{x}} = \{i | \vec{x} \in \Delta_i\}$ and call it the lattice environment of $\vec{x}$ [12].

Second, let us take as components $\rho_{ic}(\vec{x})$ and $\rho_{is}(\vec{x})$ the copies of some base functions $C(\vec{x})$ and $S(\vec{x})$, $C(\vec{x}), S(\vec{x}) \geq 0$ moved to the lattice nodes $\vec{\mu}_i$, i.e. components have the form $\rho_{ic}(\vec{x}) = C(\vec{x} - \vec{\mu}_i)$ and $\rho_{is}(\vec{x}) = S(\vec{x} - \vec{\mu}_i)$. The area Δ in which $C(\vec{x})$ or $S(\vec{x})$ are not equal to zero will be called the base carrier. This base carrier is assumed to be symmetric in the sense that it contains the origin $\vec{0} \in \Delta$ and together with each point $\vec{x} \in \Delta$ it contains $-\vec{x} \in \Delta$.

Under the assumptions made, the general recurrent algorithm (15) takes the form:

$$
\begin{aligned}
w_l^{(n+1)} &= \frac{1}{k}\sum_j^{\vec{x}_j \in \Delta_l} \frac{w_l^{(n)}C(\vec{x}_j-\vec{\mu}_l)}{\sum_i^{i\in\delta_{\vec{x}_j}} w_i^{(n)}C(\vec{x}_j-\vec{\mu}_i)+v_i^{(n)}S(\vec{x}_j-\vec{\mu}_i)}; \\
v_l^{(n+1)} &= \frac{1}{k}\sum_j^{\vec{x}_j \in \Delta_l} \frac{v_l^{(n)}S(\vec{x}_j-\vec{\mu}_l)}{\sum_i^{i\in\delta_{\vec{x}_j}} w_i^{(n)}C(\vec{x}_j-\vec{\mu}_i)+v_i^{(n)}S(\vec{x}_j-\vec{\mu}_i)};
\end{aligned}
\tag{17}
$$

Due to the changed summation limits, the system (17) becomes sparse, which, as is known, significantly reduces the requirements for memory and time resources. Indeed, if each count can belong to the intersection of not more than δ components, and each field contains, on average, k/K counts, then the amount of memory required to store intermediate data (16) will be reduced to $\sim \delta k$ values, and the number of calculations will be reduced to $\sim (\delta + 1)k$ operations, both are much less than in the general case ($\sim k \times K$).

The algorithm (17) can be further simplified if we use the fact that receptive fields of the same type intersect each other very weakly (see the above facts about the RF sets of the same type and Fig. 7 illustrating it). Assuming that there are no intersections between adjacent RFs at all, we get that each count $\vec{x}_j$ can belong to only one receptive field carrier Δ_l, i.e. $\delta_{\vec{x}_j}$ consists of only one index l of that field. So, the denominators of the right-hand sides of (17) will contain only the couple of components – the center and the surround of the given field and the system (17) is reduced to the following set of independent two-equation systems:

$$l = 1, \ldots, K :$$
$$w_l^{(n+1)} = \frac{1}{k}\sum_j^{\vec{x}_j \in \Delta_l} \frac{w_l^{(n)}C(\vec{x}_j-\vec{\mu}_l)}{w_l^{(n)}C(\vec{x}_j-\vec{\mu}_l)+v_l^{(n)}S(\vec{x}_j-\vec{\mu}_l)};$$
$$v_l^{(n+1)} = \frac{1}{k}\sum_j^{\vec{x}_j \in \Delta_l} \frac{v_l^{(n)}S(\vec{x}_j-\vec{\mu}_l)}{w_l^{(n)}C(\vec{x}_j-\vec{\mu}_l)+v_l^{(n)}S(\vec{x}_j-\vec{\mu}_l)}; \tag{18}$$

Note that by summing Eqs. (18), it is easy to obtain that at any iteration the relation $w_l^{(n)}+v_l^{(n)} = \pi_l = k_l/k$ is satisfied, where k_l is the number of counts from X_k belonging to carrier Δ_l. Thus, in (18) only one of two parameters $w_l^{(n+1)}$ or $v_l^{(n+1)}$ should be calculated, the other is obtained from the above relation automatically. It is also convenient, in view of this relation, to pass in Eqs. (18) from probabilities w_l and v_l to the conditional probabilities $\pi_{lc} = w_l/\pi_l$, $\pi_{ls} = v_l/\pi_l$, $\pi_{lc} + \pi_{ls} = 1$ introduced above, replacing the total number of counts in X_k – k by the local number of counts k_l in Δ_l:

$$l = 1, \ldots, K :$$
$$\pi_{lc}^{(n+1)} = \frac{1}{k_l}\sum_j^{\vec{x}_j \in \Delta_l} \frac{\pi_{lc}^{(n)}C(\vec{x}_j-\vec{\mu}_l)}{\pi_{lc}^{(n)}C(\vec{x}_j-\vec{\mu}_l)+\pi_{ls}^{(n)}S(\vec{x}_j-\vec{\mu}_l)};$$
$$\pi_{ls}^{(n+1)} = \frac{1}{k_l}\sum_j^{\vec{x}_j \in \Delta_l} \frac{\pi_{ls}^{(n)}S(\vec{x}_j-\vec{\mu}_l)}{\pi_{lc}^{(n)}C(\vec{x}_j-\vec{\mu}_l)+\pi_{ls}^{(n)}S(\vec{x}_j-\vec{\mu}_l)}; \tag{19}$$

Simplification (19) could be continued along in the same way if we assume that the carriers Δ_c and Δ_s of the base functions $C(\vec{x})$ and $S(\vec{x})$ do not intersect either. Considering that in this case for $\vec{x} \in \Delta_c$ we have $S(\vec{x}) = 0$ and, conversely, for $\vec{x} \in \Delta_s$ we have $C(\vec{x}) = 0$, we get that some of the terms in Eqs. (19) turn to 0, and the rest of the terms are equal to unity in the first sum for $\vec{x}_j \in \Delta_c$ and for $\vec{x}_j \in \Delta_s$ in the second. This immediately leads to simple final solutions:

$$l = 1, \ldots, K : \pi_{lc} = \frac{k_{lc}}{k_l}, \pi_{ls} = \frac{k_{ls}}{k_l}; \tag{20}$$

where k_{lc} and k_{ls} are the numbers of counts from X_k belonging to carriers of corresponding center and surround $\Delta_{lc}, \Delta_{ls} \subset \Delta_l$. Here it is interesting to note that solutions (20) do not depend at all on the forms of $C(\vec{x})$ and $S(\vec{x})$ functions (only on the shapes of their carriers).

However, solutions (20) are apparently an oversimplification of the realistic situation, so the intersection of carriers Δ_c and Δ_s should not be neglected, and (19) should be considered as the main system for finding the maximum-likelihood parameters $\vec{\theta}^{ML}$. Solutions (20) can be considered in this context as zero approximations $\pi_{lc}^{(0)}$ and $\pi_{ls}^{(0)}$, initializing algorithm. Domains Δ_c and Δ_s should be understood now as some partition of Δ: $\Delta_c \cup \Delta_s = \Delta$, $\Delta_c \cap \Delta_s = \varnothing$, determined by the forms of $C(\vec{x})$ and $S(\vec{x})$. For example, such a partition could be taken as concentric domains of Δ, the ratio of the areas of which is equal to π_c^0/π_s^0, where π_c^0 and π_s^0 are the introduced above conditional probabilities for the smoothest mixture $\rho^0(\vec{x}) = \pi_c C(\vec{x}) + \pi_s S(\vec{x})$.

The above theoretical synthesis of methods for solving the maximum likelihood (6) is illustrated below by the results of numerical experiments on encoding the image presented in Fig. 3. Recall that in the framework of the generative paradigm a code is understood as such a set of parameters $\vec{\theta}^{ML}$ (6) of the distribution density from parametric family $\mathcal{G} = \left\{ \rho(\vec{x};\vec{\theta}) \right\}$, which fits best the sample representation X_k.

The main elements of the computational scheme (19) are presented in Fig. 6. They concern, firstly, the choice of the lattice, in whose nodes $\{\vec{\mu}_i\}$, $i = 1, \ldots, K$ the receptive fields are located. The lattice is chosen to be of a rectangular type with vertical and horizontal spacing equal to d. Thus, RFs are located at the vertices of square lattice cells with dimensions $d \times d$. The base carrier Δ is also chosen to be square with dimensions $d \times d$, which, when copied to the nodes $\{\vec{\mu}_i\}$, ensures dense covering without overlapping of the surface Ω by the set $\{\Delta_i\}$ (see Fig. 6A). The size of the base carrier $D = d$ is assumed to be an integer and is equal to the number of jots of the sensitive surface Ω along any of its sides. In other words, hereinafter, the size of an individual jot (pixel) is taken as unity. The number of cells in the rows/columns of the lattice is assumed to be equal to l, so the total number of RFs is $K = l^2$, and the total number of jots is $N = (ld)^2$.

The second main choice concerns the forms of components $\rho_{ic}(\vec{x})$ and $\rho_{is}(\vec{x})$, which are the moved copies of base functions $C(\vec{x})$ and $S(\vec{x})$, $C(\vec{x}), S(\vec{x}) \geq 0$. The basic functions are chosen to be axisymmetric with respect to the origin of coordinates. It allows them to be specified using one-dimensional profiles depending on $r = \|\vec{x}\|$, which in this case are selected in the form (see Fig. 6B)):

$$C(r) = \frac{1}{\pi\sigma^2}\exp\{-r^2/\sigma^2\},$$
$$S(r) = \frac{1}{d^2-\pi\sigma^2}(1 - \exp\{-r^2/\sigma^2\}); \tag{21}$$

where σ is the characteristic size of the center of the receptive field, which could be adjusted in the calculations with respect to the field size d, so that the ratio $\gamma = \sigma/d$ is an instrumental parameter of the method. Looking ahead, we note that the best results in the experiments were obtained at $\gamma \approx 2$. The choice of profiles (21) was motivated by the consideration that there are $\pi_c^0 = \pi\sigma^2/d^2$ and $\pi_s^0 = (d^2-\pi\sigma^2)/d^2$, which for a smooth mixture $\rho^0(\vec{x}) = \pi_c^0 C(\vec{x}) + \pi_s^0 S(\vec{x})$ give exactly a uniform distribution $\Pi(\vec{x})$. This distribution, together with the difference (center-surround selective operator) $D(\vec{x}) = C(\vec{x}) - S(\vec{x})$, are shown in Fig. 6C.

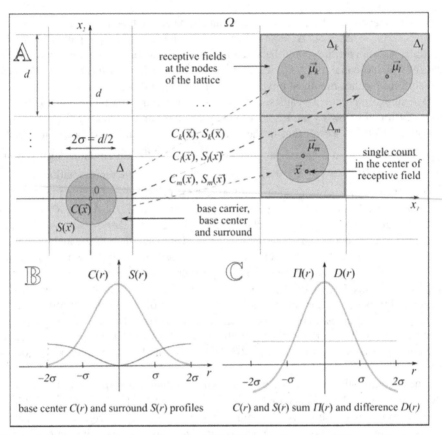

Fig. 6. Computational scheme of algorithm (19). A) The outline of the lattice and the locations of receptive fields on Ω with the structure of the basic center/surround domains. B) Profiles of the center $C(\vec{x})$ and surround $S(\vec{x})$ (21). C) profiles of the smoothest mixture $\rho^0(\vec{x})$ (uniform distribution $\Pi(\vec{x})$) and difference $D(\vec{x})$ of the center and surround.

Following are the encoding results of the "GRAY_R02_0600x0600_093" image sampling representations, shown in the Fig. 3. The results are presented as screenshots of a computer application specially designed for the current investigation. The application was written on the platform.NET Framework 4.6.1 of integrated development environment Microsoft Visual Studio Community 2019.

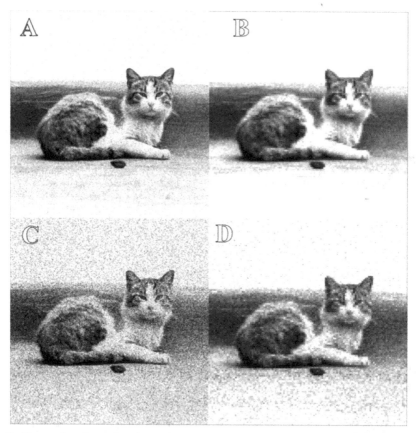

Fig. 7. The results of encoding the sampling representations for different sizes k, shown in Fig. 3. A) Sampling representation for $k = 5000000$ counts (Fig. 3.D)). B) Encoding the sampling representation A. C) Sampling representation for $k = 1000000$ counts (Fig. 3.C)). D) Encoding the sampling representation C.

Later on if it is not explicitly stated, the following parameters of the algorithm (19) have been used: the size of the sampling representation (number of counts) $k = 1000000$, the size of the RF size $d = 10$, the number of cells in the rows/columns of the lattice $l = 120$ ($K = 14400$), center/RF size ratio $\gamma = 4$, number of iterations in (19) $n + 1 = 10$.

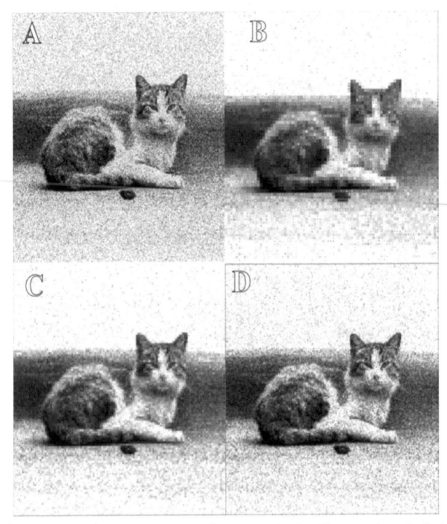

Fig. 8. The results of encoding the sampling representation for different numbers of cells in lattice rows/columns. A) Sampling representation for $k = 1000000$ counts (Fig. 3C)). B) Encoding sampling representation A by a lattice of $l = 80$ cells in row. C) Encoding sampling representation A by a lattice of $l = 120$ cells in row. D) Encoding sampling representation A by a lattice of $l = 200$ cells in row.

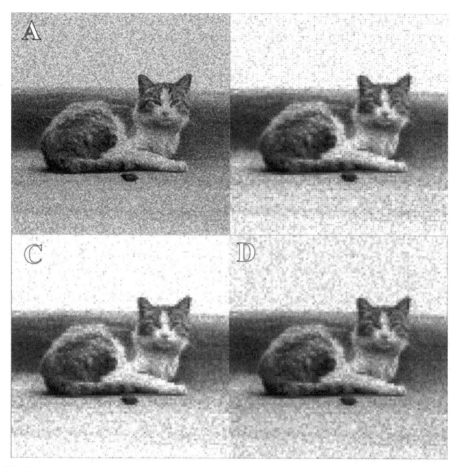

Fig. 9. The results of encoding the sampling representation for different parameters $\gamma = \sigma/d$ of center to field size ratio. A) Sampling representation for $k = 1000000$ counts (Fig. 3.C)). B) Encoding sampling representation A by a ratio $\gamma = 3$. C) Encoding sampling representation A by a ratio $\gamma = 4$. D) Encoding sampling representation A by a ratio $\gamma = 6$.

4 Conclusions

The synthesized method of perceptual image coding showed good characteristics in many respects. Firstly, this is a fast method focused on big data. With the number of calculations in several iterations on sampling representations of several million counts, code calculation requires only a fraction of a second for a modern personal computer with average performance. Secondly, a medium-sized image "GRAY_R02_0600x0600_093" (600 × 600 pixels, gray, 8-bit color depth, 144 KB on disk) being compressed to a set of $K = 2 \times 100 \times 100 = 20000$ mixture weights (20 KB) has nevertheless sufficient visual quality (see Figs. 7–9) for understanding the image, recognizing objects on it, etc.

Note that the achieved characteristics are only the first results obtained on the way of modeling the mechanisms of visual perception, and so far only at the level of retinal functions. In this regard, we express the cautious hope that the potential of the proposed approach can be revealed to an even greater extent in subsequent studies.

References

1. Fossum, E.: The invention of CMOS image sensors: a camera in every pocket. In: 2020 Pan Pacific Microel. Symp., pp. 1–6. (2020). https://doi.org/10.23919/PanPacific48324.2020.9059308

2. Gigajot Announces the World's Highest Resolution Photon Counting Sensor, Press release: https://www.prnewswire.com/news-releases/gigajot-announces-the-worlds-highest-resolution-photon-counting-sensor-301516410.html. Accessed 01 June 2022

3. Photon-Counting Image Sensors. Fossum, E. R., Teranishi, N., et al (eds.). MDPI (2017)

4. Sejnowski, T.J.: The unreasonable effectiveness of deep learning in artificial intelligence. In: Proceedings of the National Academy of Sciences, 117(48), pp. 30033–30038. (2020). https://doi.org/10.1073/pnas.1907373117

5. Bishop, C.M., Lasserre, J.: Generative or discriminative? getting the best of both worlds. In: Bernardo, J.M. et al. (eds.) Bayesian Statistics 8. International Society for Bayesian Analysis, pp. 3–24. Oxford University Pres (2007)

6. Goodfellow, I., et al.: Generative adversarial networks. Commun. ACM **63**(11), 139–144 (2020). https://doi.org/10.1145/3422622

7. Kingma, D.P., Welling, M.: Auto-Encoding Variational Bayes. CoRR /1312.6114 (2014)

8. Hinton, G.E., Osindero, S., The, Y.-W.: A fast-learning algorithm for deep belief nets. Neural Comput. **18**(7), 1527–1554 (2006). https://doi.org/10.1162/neco.2006.18.7.1527

9. Hassabis, D., Kumaran, D., Summerfield, C., Botvinick, M.: Neuroscience-inspired artificial intelligence. Neuron **95**(2), 245–258 (2017). https://doi.org/10.1016/j.neuron.2017.06.011

10. Stroustrup, B.: What is object-oriented programming? IEEE Softw. **5**(3), 10–20 (1988). https://doi.org/10.1109/52.2020

11. Kershner, V., Antsiperov, V.: Image coding by samples of counts as an imitation of the light detection by the retina. In: Proceedings of the 11th International Conference on Pattern Recognition Applications and Methods - ICPRAM, pp. 41–50. (2022). https://doi.org/10.5220/0010836800003122

12. Antsiperov, V.E.: Representation of images by the optimal lattice partitions of random counts. Pattern Recogn. Image Anal. **31**(3), 381–393 (2021). https://doi.org/10.1134/S105466182103 0044

13. Scott, D.W.: Multivariate density estimation: theory, practice, and visualization, 2nd edn. Wiley, Hoboken, New Jersey (2015)

14. McLachlan, G.J., Lee, S.X., Rathnayake, S.I.: Finite mixture models. Annual Rev. Stat. Appl. **6**(1), 355–378 (2019). https://doi.org/10.1146/annurev-statistics-031017-100325

15. Aldrich, J.R.A.: Fisher and the making of maximum likelihood 1912–1922. Stat. Sci. **12**(3), 162–176 (1997)

16. Streit, R.L.: Poisson Point Processes. Imaging, Tracking and Sensing. Springer, New York (2010)

17. Fox, M.: Quantum Optics: An Introduction. Oxford University Press, Oxford, New York (2006)

18. Bertero, M., Boccacci, P., Desidera, G., Vicidomini, G.: Image deblurring with Poisson data: from cells to galaxies. Inverse Prob. **25**(12), 123006 (2009). https://doi.org/10.1088/0266-5611/25/12/123006

19. Bateman, H., Erdelyi, A.: Higher Transcendental Functions, vol. 2. McGraw-Hill, New York (1953)

20. Bengio, Y., Courville, A., Vincent, P.: Representation learning: a review and new perspectives. IEEE Trans. Pattern Anal. Mach. Intell. **35**(8), 1798–1828 (2013). https://doi.org/10.1109/TPAMI.2013.50
21. Robert, C.P., Casella, G.: Monte Carlo Statistical Methods, 2nd ed. Springer-Verlag, New York (2004). https://doi.org/10.1007/978-1-4757-4145-2
22. Asuni, N., Giachetti, A.: TESTIMAGES: a large data archive for display and algorithm testing. J. Graph. Tools **17**(4), 113–125 (2015). https://doi.org/10.1080/2165347X.2015.1024298
23. Jayant, N.: Signal compression. Int. J. High-Speed Electron. Syst. **8**(1), 1–12 (1997). https://doi.org/10.1142/s0129156497000020
24. Jayant, N, Johnston, J., Safranek, R.: Perceptual coding of images. In: Proceedings of SPIE, 1913(1), pp. 168–178 (1993). https://doi.org/10.1117/12.152691
25. Blau, Y., Michaeli, T.: Rethinking lossy compression: the rate-distortion-perception tradeoff. In: Proceedings of the 36th International Conference on Machine Learning, PMLR 2019, 97, pp. 675–685 (2019)
26. Schiller, P.H., Tehovnik, E.J. Vision and the Visual System. Oxford University Press, Oxford (2015). https://doi.org/10.1093/acprof:oso/9780199936533.001.0001
27. Kuffler, S.W.: Discharge patterns and functional organization of mammalian retina. J. Neurophysiol. **16**(1), 37–68 (1953). https://doi.org/10.1152/jn.1953.16.1.37
28. Kling, A., Field, G.D., Brainard, D.H., Chichilnisky, E.J.: Probing computation in the primate visual system at single-cone resolution. Annu. Rev. Neurosci. **42**(1), 169–186 (2019). https://doi.org/10.1146/annurev-neuro-070918-050233
29. Bear, M.F., Connors, B.W., Paradiso, M.A.: Neuroscience: exploring the brain, 4th edn. Wolters Kluwer, Philadelphia (2016)
30. Gauthier, J.L., Field, G.D., et al.: Receptive fields in primate retina are coordinated to sample visual space more uniformly. PLoS Biol. **7**(4), e1000063 (2009). https://doi.org/10.1371/journal.pbio.1000063
31. Field, G.D., Gauthier, J.L., et al.: Functional connectivity in the retina at the resolution of photoreceptors. Nature **467**(7316), 673–677 (2010). https://doi.org/10.1038/nature09424
32. Bishop, C.M., Lasserre, J. Generative or Discriminative? Getting the Best of Both Worlds. Bayesian Statistics 8, pp. 3–24. Oxford University Press, London (2007)
33. Mallat, S. A theory for multiresolution signal decomposition: the wavelet representation. IEEE Trans. Pattern Recogn. Mach. Intell. **11**(7), pp. 674–693 (1989). https://doi.org/10.1109/34.192463
34. Kruizinga, P., Petkov, N.: Computational model of dot-pattern selective cells. Biol. Cybern. **83**(4), 313–325 (2000). https://doi.org/10.1007/s004220000153
35. Berinde, V.: Iterative Approximation of Fixed Points, 2nd edn. Springer, Berlin (2007)
36. McLachlan, G.J., Krishnan, T.: The EM Algorithm and Extensions, 2nd edn. Wiley, Hoboken, New York (2007)
37. Nguyen, H.D., Forbes, F., McLachlan, G.J.: Mini-batch learning of exponential family finite mixture models. Stat. Comput. **30**(4), 731–748 (2020). https://doi.org/10.1007/s11222-019-09919-4

Gesture Recognition and Multi-modal Fusion on a New Hand Gesture Dataset

Monika Schak$^{(\boxtimes)}$ and Alexander Gepperth

Fulda University of Applied Sciences, 36037 Fulda, Germany
{monika.schak,alexander.gepperth}@cs.hs-fulda.de

Abstract. We present a baseline for gesture recognition using state-of-the-art sequence classifiers on a new freely available multi-modal dataset of free-hand gestures. The dataset consists of roughly 100,000 samples, grouped into six classes of typical and easy-to-learn hand gestures. The dataset was recorded using two independent sensors, allowing for experiments on multi-modal data fusion at several depth levels and allowing research on multi-modal fusion for early, intermediate, and late fusion techniques. Since the whole dataset was recorded by a single person we ensure a very high quality of data with little to no risk for incorrectly performed gestures. We show the results of our experiments on unimodal sequence classification using a LSTM as well as a CNN classifier. We also show that multi-modal fusion of all four modalities results in higher precision using late-fusion of the output layer of an LSTM classifier trained on a single modality. Finally, we demonstrate that it is possible to perform live gesture classification using an LSTM-based gesture classifier, showing that generalization to other persons performing the gestures is high.

Keywords: Hand gestures · Dataset · Multi-modal data · Data fusion · Sequence classification · Gesture recognition

1 Introduction

We present a freely available multi-modal dataset of freehand gestures that can be used for research on sequence classification, multi-modal fusion, or other domains in Human-Computer-Interaction and Machine Learning. Hand gesture recognition is widely used as a natural way of non-verbal communication. In general, there are two main applications for hand gesture recognition: Communication, e.g. in sign language recognition, and manipulation, e.g. controlling a robot or other technical device. Another typical application is controlling and communicating with a virtual environment [8].

Large, multi-modal, and reliable datasets are needed for modern deep learning techniques to perform at a sufficiently high level. Important characteristics of a good dataset with training data are the number of gesture classes, the number of samples per gesture class, and the number of distinct modalities from (different) sensors. (cf. [24]).

We present results for baseline experiments to show that the dataset can be used to train state-of-the-art machine learning models and can achieve very high prediction rates. For this, we use Long Short-Term Memory (LSTM) networks as well as deep

© Springer Nature Switzerland AG 2023
M. De Marsico et al. (Eds.): ICPRAM 2021/2022, LNCS 13822, pp. 76–97, 2023.
https://doi.org/10.1007/978-3-031-24538-1_4

Convolution Neural Networks (CNN). LSTM networks [9] are widely used recurrent neural networks with feedback connections to process sequences instead of just single data points. CNN are feed-forward neural networks consisting of fully-connected layers, pooling layers, and normalization layers. They are most common in image classification.

When working with multiple modalities, each sensor produces a separate data stream, also called sensory modality. Each sensory modality contains unique and independent information. But since all sensors observe the same situation – in our scenario the same hand gesture – the information from all sensors at least partially correlates. The goal of multi-modal fusion is to exploit that correlation to obtain more precise and reliable observations. In general, sensor data can be fused at three stages. Early fusion approaches combine data from sensors without preprocessing or after features have been extracted from raw data and use collaborative representation classifiers [14]. Late fusion strategies often combine the output score provided by multiple classifiers each trained on a single modality [28], i.e. by transforming the output to a probability score by a softmax layer and combining it by sum rule, product rule, or max rule.

This paper is an extended version of our previous work [25]. In this version, we conducted additional experiments using CNNs for uni-modal classification. We also more thoroughly explained the reasoning behind our choices regarding the single-user approach as well as the selection of our gesture classes.

2   Related Work

2.1   Hand Gesture Datasets

Several hand gesture datasets have been made available to researchers in recent years. Regardless, we find that there are no publicly available datasets that include a large enough number of gesture samples needed to suit the high requirement of machine learning methods for an extensive dataset for training. On the other hand, we require a dataset with a reasonable number of modalities recorded from independent sensors since present-day sensors are increasingly cheap and universally available and we find that gesture recognition can highly profit when including information from several modalities.

The SHGD dataset [13] consists of 15 gesture classes recorded from 27 persons with 96 sequence samples per class, resulting in a total size of 4,500 gesture samples. It only contains depth data recorded by an RGB-D camera.

A multi-modal dataset is presented in [16]. It consists of 10 gesture classes recorded from 14 persons with 140 sequence samples per class, resulting in a total size of 1,400 gesture samples. It contains depth data recorded by an RGB-D camera and data from a LeapMotion sensor.

In the next dataset [19], the approach is to render gesture samples using an advanced computer graphics pipeline instead of recording them. The dataset consists of 11 gesture classes with about 3,000 sequence samples per class, totaling 35,200 gesture samples. It only contains depth data.

The Cambridge dataset [12] contains 10 gesture classes recorded from two people with about 100 sequence samples per class, resulting in a total size of 1,000 gesture

Table 1. Comparison of the MMHG dataset with other hand-gesture datasets provided in literature. (Source: [25].)

Dataset	Classes	Samples/Class	Persons	Total samples	Modalities
SHGD [13]	15	96	27	4,500	Depth
Cambridge dataset [12]	10	100	2	1,000	RGB
n.A. [16]	10	100	14	1,400	Depth, Motion
IsoGD [30]	249	190	21	50,000	RGB, Depth
EgoGesture [32]	83	300	50	24,000	RGB, Depth
SKIG [15]	10	360	6	1,080	RGB, Depth
ChaLearn [6]	20	390	27	13,900	Audio, RGB, Depth
n.A. [18]	11	3,000	–	35,200	Rendered Depth
MMHG (this paper)	6	≈13,300	1	79,881	RGB, Depth, Motion, Audio

samples. The Sheffield Kinect Gesture Dataset [15] contains 10 gesture classes recorded from six persons with 360 sequence samples per class, resulting in a total size of 1,080 gesture samples. It also only contains depth data recorded by an RGB-D camera.

One of the first large-scale hand gesture datasets is the ChaLearn-2013 dataset [6], which consists of 20 gesture classes recorded from 27 persons with an average of 360 gesture samples per class, resulting in a total size of roughly 14,000 sequence samples. It contains audio, RGB, and depth modality.

The IsoGD dataset [29] is even larger but only includes an RGB and a depth modality. It contains about 50,000 gesture samples, grouped into 249 gesture classes with an average of 190 gesture samples per class, and it was recorded by 21 persons.

The next dataset [32] is of similar size and also includes only an RGB and a depth modality, but it is egocentric and recorded from a head-mounted camera. The dataset consists of 83 gesture classes with about 300 sequence samples per class, totaling about 24,000 gesture samples.

Table 1 shows a comparison of the MMHG dataset with the hand-gesture datasets introduced in this section.

2.2 Multi-modal Fusion

Research in the field of psychology and neurophysiology [1,2] shows multi-sensory fusion to be a common concept. This means, that the human brain is capable of probabilistically combining different modalities [5,7].

Similar concepts are available in the field of multi-modal gesture or activity recognition. There is not the one right way to carry out multi-modal fusion, but there are several possibilities that have been used in experiments in recent years. Each possibility has its own advantages and disadvantages depending on the data used and tasks at hand.

In general, multi-modal fusion techniques can be clustered into three categories: early fusion, intermediate fusion, and late fusion. Early fusion describes that data from different sensory modalities are combined either before any preprocessing steps [14] or after features have been extracted from raw data [4] by using a collaborative representation classifier. The fused data is then passed onto a machine learning model.

In late fusion techniques, there usually are multiple machine learning models for single modalities which are fused at a later stage, very often right before output score prediction. In [27], single modality-based classifiers each provide an output score and the final score is produced by searching the maximum. Another possibility is to use a collaborative representation classifier to combine classification outcomes of different modalities [4]. A very easy and therefore commonly used late-fusion technique is softmax score fusion [11]. Here, multiple classifier outputs are seen as probability scores by a softmax layer. Afterward, they are combined using either the sum rule, the product rule, or the max rule.

Intermediate fusion happens between data level and output level. A possible technique is feature fusion [10]. Features that are the output from fully connected layers are combined and then forwarded to a linear support vector machine or any other classifier.

2.3 Contribution

This work, like the original work [25], has its focus on describing our Multi-Modal Hand Gesture Dataset. Almost 80,000 samples with over 13,000 samples per gesture class in four modalities. There are six gesture classes that are all supposed to be easy to perform by the user but also specifically chosen to make the dataset beneficial for research on multi-modal fusion. All samples have been recorded by just one very well trained and instructed user, therefore the dataset does not contain corrupted data samples. The recording and preprocessing steps were carefully designed to ensure high quality. Thus, the dataset is perfectly suited for machine learning.

Additionally, we present experiments that prove the consistency of the dataset. The first set of experiments shows that plausible classification accuracies can be achieved on each of the four modalities when trained on two state-of-the-art sequence classification models: LSTM networks and CNNs. The second set of experiments shows that even with relatively simple late multi-modal fusion approaches it is possible to improve the classification accuracies achieved by networks trained on uni-modal data.

In Sect. 6 we introduce an implementation based on the Robot Operating System and the results for our experiments to determine the generalization capabilities of our dataset to other people. This shows that although only a single user recorded all data samples for the dataset, it is still possible to train a gesture classifier capable of correctly classifying gestures performed by other users.

3   Dataset

In this paper, we present the Multi-Modal Hand Gesture Dataset (MMHGD). It is a large-scale dataset with only six classes but a high number of samples for each class. Each sample consists of modalities from an RGB and 3D camera, a microphone, and an acceleration sensor.

All gesture samples are recorded and performed by just one single person. This is an unusual choice but since only a well-instructed person performs all gestures, there will be no incorrectly performed gesture samples in the dataset. Thus, we ensure a high quality of gesture recordings and little to no corrupted data.

The dataset contains about 13,300 recordings of each of the six classes. Therefore, it contains a total of almost 80,000 samples. Table 2 shows the exact distribution of each class. Each gesture sample consists of data from a two-second window. The dataset (raw as well as preprocessed data) can be downloaded at http://data.informatik.hs-fulda.de.

Table 2. Distribution of the six gesture classes in the MMHG dataset.

Class	C_1	C_2	C_3	C_4	C_5	C_6	Total
Samples	13,440	13,410	13,228	13,233	13,308	13,262	**79,881**

Although the gestures were recorded with varying background and lightning, we used a fixed setup to ensure that each sample is recorded within a predefined distance of 0.5 m to 0.75 m from the camera. Thus, simplifying the preprocessing step (Fig. 1).

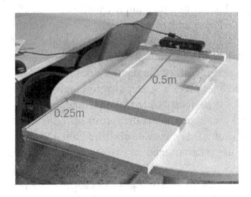

Fig. 1. The setup used for recording the gesture samples for the MMHG dataset, ensuring a fixed distance to the camera (Source: [25]).

During the time of recording, the user is told which gesture to perform and the recorded sample is immediately assigned the correct class label. Each recording therefore consists of RGB images, 3D point clouds, an mp3 file, the acceleration data and the correct class label. We also have preprocessed data available to immediately use for training and testing. The preprocessing step happens independently of the recording and is described in Sect. 3.3.

3.1 Gesture Classes

We use easy-to-use gesture classes that are commonly used in human-machine-interactions. The gesture classes are specifically designed to offer challenging tasks for multi-modal fusion. Therefore, some of the classes rely heavily on fusing different modalities for correct classification. For example, the last two classes – Snap Once and

Snap Twice – do not differ much in their motion. Thus, using only visual modalities will probably not lead to very good classification results. However, fusing those modalities with the audio modality will make them easily distinguishable.

Thumbs Up. (Class 1, denoted as C_1): The first class describes a thumbs-up gesture, which is a very typical gesture to show approval or agreement. The gesture starts with a closed fist in front of the camera. During the gesture, the thumb is extended upwards. There is no distinctive sound for this gesture and it is rather stationary with very little movement of the whole hand. Four frames of the RGB modality of one gesture sample are shown in Fig. 2 exemplarily.

Fig. 2. Four frames of the RGB modality of one Thumbs Up gesture sample.

Thumbs Down. (Class 2, denoted as C_2): The second class is similar to the first class as it describes a thumbs-down gesture. This gesture is commonly used to show rejection or disapproval. This gesture also starts with a closed fist in front of the camera, but this time the thumb is extended downwards. Just like the thumbs-up gesture class, there is no distinctive sound. It shows more movement since the fist is tilted forwards while the thumb is extended downwards. Figure 3 shows four frames of the RGB modality of one gesture sample as an example.

Fig. 3. Four frames of the RGB modality of one Thumbs Down gesture sample.

Swipe Left. (Class 3, denoted as C_3): The third class shows a swiping gesture of the whole hand in a horizontal direction from the right side to the left. Therefore, it is a dynamic gesture. It can be used to switch or forward to the next element or move an object from right to left. Again, this gesture does not have a distinctive sound. Four example frames from the RGB modality of one gesture sample can be seen in Fig. 4.

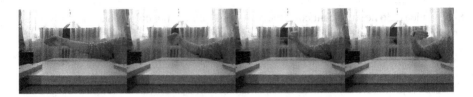

Fig. 4. Four frames of the RGB modality of one Swipe Left gesture sample.

Swipe Right. (Class 4, denoted as C_4): The fourth class is a swipe right gesture, similar to the previous gesture class. It also is a dynamic hand gesture that describes a swiping gesture of the whole hand in a horizontal direction from the left side to the right. It can be used to switch or rewind to the previous element or move an object from left to right. Unsurprisingly, this gesture also does not have a distinctive sound. Figure 5 shows four frames from the RGB modality of one gesture sample.

Fig. 5. Four frames of the RGB modality of one Swipe Right gesture sample.

Snap Once. (Class 5, denoted as C_5): The fifth class again is a rather stationary gesture, where the hand remains almost still while only two fingers show motion. For this gesture, the middle finger pushes hard against the thumb and then gets released so fast that it creates a snapping sound. Therefore, this gesture class has a distinctive sound. Four frames from the RGB modality of one gesture sample from this class are shown in Fig. 6.

Fig. 6. Four frames of the RGB modality of one Snap Once gesture sample.

Snap Twice. (Class 6, denoted as C_6): The sixth gesture class is very similar to the previous class and describes two snaps. Accordingly, the hand remains almost still while only two fingers are moving. For this gesture, the middle finger and the thumb create two snapping sounds consecutively. This gesture class also has a distinctive sound. Figure 7 shows four frames of the RGB modality of one gesture sample as an example.

Fig. 7. Four frames of the RGB modality of one Snap Twice gesture sample.

3.2 Modalities

RGB. The sensory modality of RGB images is recorded with an Orbbec Astra 3D sensor. It outputs a stream of images at a resolution of 800×600 pixels with a frequency of 30 fps. To reduce the load and still be able to process all modalities at the same time, we save images at a frequency of 6 fps. This leads to twelve images for every gesture sample.

3D. The sensory modality of 3D images is also recorded with an Orbbec Astra 3D sensor. The 3D camera outputs a stream of depth images with a size of 640×480 pixels. Those images are converted to point clouds before being saved for further preprocessing. Due to performance reasons and to get matching RGB images to every 3D image, we also save point clouds at a frequency of 6 fps. Thus, receiving twelve point clouds for every gesture sample.

Audio. The audio data is also provided by the Orbbec Astra 3D sensor. The sensor has a sensitivity of 30 dB and works with audio between 20 Hz and 16 kHz. We save the raw wave data for the entire two-second window for every gesture sample.

Acceleration. The sensory modality of the acceleration data is recorded using an acceleration sensor (BWT901CL from Bitmotion) attached to the users right wrist. The acceleration sensor offers 9-axis: acceleration data in three axis, yaw rates also in three axis, gyroscopic measurements, and magnetic field measurements. The sensor has a frequency of 200 Hz, which means we get 400 measurements for each gesture sample.

3.3 Preprocessing

RGB. To reduce the computational costs and remove unnecessary overload, we crop the 800×600 px RGB images to the part where the hand is visible. Examples for the original RGB images can be seen in Figs. 2, 3, 4, 5, 6 and 7. Since we always perform the hand gesture in a predefined area in front of the camera, the complexity of this step is reduced: The hand is always in the same area in the RGB image for every frame for every gesture. Therefore, we do not have to perform object detection on every single image but instead can define the area to which to crop and it will work for all images. Afterward, we scale the cropped image to 72×48 pixels and then calculate the Histogram of Oriented Gradients (HOG) [17,31] descriptor. For this, we use the OpenCV implementation with default parameters. The only parameters we set ourselves are the cell size, which we set to 8×8 pixels, and the block size, which we set to 16×16 pixels. The calculated HoG descriptor has 756 entries. The resulting NumPy array consists of twelve HoG descriptors for every gesture sample and thus has a shape of $(N, 12, 756)$.

3D. During the recording phase we store twelve point clouds for every gesture sample. One exemplary frame from a sample of the Thumbs Up gesture is shown in Fig. 8a. Each of these point clouds goes through the same three steps during the preprocessing phase. The first step is downsampling the point cloud to reduce the size and computational costs of the following steps. We begin by removing measurement errors by deleting all points where one or more of the x-, y-, or z-value is not a number (NaN), then we perform downsampling using the 3D-voxel grid technique. Using this downsampled point cloud, we perform conditional removal to delete all points that are outside of our predefined volume of interest. Again, this can be done using the same volume of interest for every single gesture frame since we ensured that the gesture is always performed in the same area in front of the camera. The result is a point cloud of just the hand without any background data. In the second step, we infer surface normals by using approximation and use those to calculate Point Feature Histograms (PFH) [22,23] in the third step. With PFHs, we are able to receive a descriptor with the same size for every point cloud - although they have a high variability in size. This is important since machine learning models often require a fixed input size. According to [21], we randomly select two surface normals and compute the "four values based on the length and relative orientation of the surface normals" [25]. By dividing each value into five intervals, we receive 625 possible discrete values, which then get normalized. The resulting histogram consisting of 625 dimensions is "able to feasibly characterize the hand and fingers" [25]. Figure 8a shows a single point cloud (frame) of the Thumbs Up gesture class, while Fig. 8b shows the corresponding PFH. The resulting NumPy array has a shape of $(N, 12, 625)$.

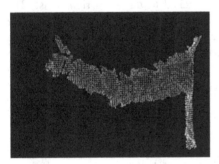

(a) Example of one point cloud of a sample for the gesture class Thumbs Up.

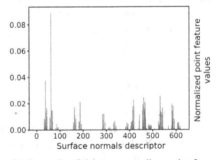

(b) Example of the corresponding point feature histogram to Figure 8a.

Fig. 8. Example of one frame of a gesture sample from the gesture class Thumbs Up (C_1) before (a) and after (b) preprocessing (Source: [25]).

Audio. We downsample the raw wave data to a frequency of 8,000 Hz and ensure an equal length by using zero-padding with a randomly picked offset. Afterwards, Short-Time Fourier Transform (STFT) [3,20] with the following parameters is performed: a window of 455 data points with an overlap of 420 data points. Using STFT we can visualize the frequency information in 2D, more precisely the change of frequency during certain time frames. The result is a NumPy array with a shape of $(N, 182, 181)$. Exemplarily, STFT data for a gesture sample with no distinct audio (Swipe Left, C_3) is shown in Fig. 9a. Figure 9b shows the STFT data for a gesture sample with one snap, while Fig. 9c shows the STFT data for a gesture sample with two snaps. While Fig. 9a

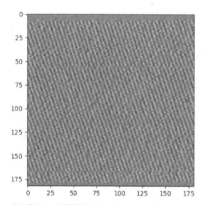

(a) Plotted STFT data for a gesture sample without distinct audio of gesture class Swipe Left (C_3).

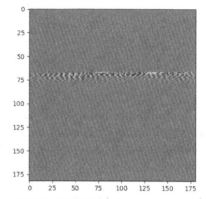

(b) Plotted STFT data for a gesture sample with distinct audio of gesture class Snap Once (C_5)

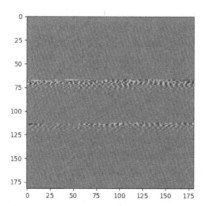

(c) Plotted STFT data for a gesture sample with distinct audio of gesture class Snap Twice (C_6).

Fig. 9. Examples of the STFT data of gesture classes without (a) and with (b, c) distinct audio (Source: [25]).

shows no distinct change in frequency, Figs. 9b and 9c show distinct changes in frequency during one or two time frames respectively.

3.4 Acceleration

The acceleration data is rather noisy. Thus, we calculate statistical values over the 20 tuples from each 200-millisecond window.

$$\bar{x} = \frac{1}{N}\left(\sum_{i=1}^{N} x_i\right) = \frac{x_1 + x_2 + \cdots + x_N}{N} \tag{1}$$

$$\mathrm{Var}(x) = \frac{1}{N}\sum_{i=1}^{N}(x_i - \bar{x})^2 \tag{2}$$

$$S(x) = \sqrt{\frac{1}{N-1}\sum_{i=1}^{N}(x_i - \bar{x})^2} \tag{3}$$

In our two-second window for every gesture sample, we receive a total of ten descriptors. Each descriptor contains the mean (cf. Eq. 1), the variance (cf. Eq. 2) and the standard deviation (cf. Eq. 3) for each of the six axes. The resulting NumPy array has a shape of $(N, 10, 3, 6)$.

4 Unimodal Classification

To show that every single modality of the MMHG dataset can be used to train state-of-the-art machine learning models to be able to perform with high gesture classification accuracies, we provide results of unimodal classification experiments with different architectures. We conducted experiments on the modalities with sequential data (RGB, 3D, and acceleration) using LSTM networks. Then, we conducted experiments on all four modalities using CNNs.

4.1 LSTM

We provide results of our experiments (cf. [25]) using LSTM networks for the RGB, 3D, and acceleration modalities, since after preprocessing those three modalities consist of sequential data, while the audio data is only one 2D plot per gesture sample. LSTM networks [9] are recurrent neural networks capable of learning dependencies over time (long-term). They are often used in sequence classification since due to the feedback connections they are able to process sequences of data instead of single data points such as in images.

We randomly select 20% of our dataset to use as a test set and train the LSTM network on the remaining 80%. Preliminary experiments were used to determine network parameters for every modality that result in the highest classification accuracies.

The gesture classification accuracy is the fraction of correct predictions compared to all predictions, as shown in Eq. 4. Precision defines the proportion of correct positive classifications compared to all positive classifications for a class, as shown in Eq. 5. Recall defines how many samples were classified correctly compared to all predictions of that class, as shown in Eq. 6. The F1-score is the harmonic mean between precision and recall, as shown in Eq. 7. (tp = true positive, tn = true negative, fp = false positive, fn = false negative)

$$\text{Accuracy} = \frac{\text{tp} + \text{tn}}{\text{tp} + \text{tn} + \text{fp} + \text{fn}} \tag{4}$$

$$\text{Precision} = \frac{\text{tp}}{\text{tp} + \text{fp}} \tag{5}$$

$$\text{Recall} = \frac{\text{tp}}{\text{tp} + \text{fn}} \tag{6}$$

$$F_1 = \frac{2}{\text{recall}^{-1} + \text{precision}^{-1}} = \frac{2 \cdot \text{tp}}{2 \cdot \text{tp} + \text{fp} + \text{fn}} \tag{7}$$

RGB. For the RGB modality, we use an LSTM network with $S = 200$ cells each on $L = 2$ hidden layers and train it in $I = 3,000$ iterations with a batch size of $b = 250$. After training the network on our training set with those parameters, we achieve an average gesture classification accuracy on the test set of 85.55%. Table 3a shows the confusion matrix for the unimodal gesture classification, while Table 3b shows the precision, recall and F1-score for every class.

Table 3. Results for the unimodal classification on the RGB modality using an LSTM network (Source: [25]).

Predicted class $C_{[1-6]}$					
2662	26	0	0	0	0
0	2335	90	38	26	193
0	165	2430	26	25	0
0	76	64	2430	76	0
0	166	0	25	1806	664
0	318	0	38	293	2004

(rows labelled Target $C_{[1-6]}$)

(a) Confusion matrix for an LSTM network trained on RGB data.

Class	Precision	Recall	F1-Score
C_1	1.00	0.99	0.99
C_2	0.75	0.86	0.80
C_3	0.94	0.92	0.93
C_4	0.95	0.92	0.93
C_5	0.80	0.67	0.73
C_6	0.69	0.74	0.72

(b) Classification report for an LSTM network trained on RGB data.

3D. For the 3D modality, we use an LSTM network with $S = 250$ cells each on $L = 2$ hidden layers and train it in $I = 5,000$ iterations with a batch size of $b = 1,000$. After training the network on our training set with those parameters, we achieve an average gesture classification accuracy on the test set of 93.43%. Table 4a shows the confusion matrix for the unimodal gesture classification, while Table 4b shows the precision, recall and F1-score for every class.

Table 4. Results for the unimodal classification on the 3D modality using an LSTM network (Source: [25]).

Predicted class $C_{[1-6]}$							Class	Precision	Recall	F1-Score
2688	0	0	0	0	0		C_1	1.00	1.00	1.00
2	2667	4	8	1	0		C_2	1.00	0.99	1.00
0	4	2613	29	0	0		C_3	0.99	0.99	0.99
0	0	16	2627	3	0		C_4	0.98	0.99	0.99
1	0	0	0	2350	310		C_5	0.74	0.86	0.80
0	0	1	0	670	1982		C_6	0.84	0.70	0.77

(a) Confusion matrix for an LSTM network trained on 3D data.

(b) Classification report for an LSTM network trained on 3D data.

Acceleration. For the acceleration modality, we use an LSTM network with $S = 250$ cells each on $L = 5$ hidden layers and train it in $I = 1,000$ iterations with a batch size of $b = 500$. After training the network on our training set with those parameters, we achieve an average gesture classification accuracy on the test set of 83.66%. Table 5a shows the confusion matrix for the unimodal gesture classification, while Table 5b shows the precision, recall and F1-score for every class.

Table 5. Results for the unimodal classification on the acceleration modality using an LSTM network (Source: [25]).

Predicted class $C_{[1-6]}$						Class	Precision	Recall	F1-Score
2571	25	20	12	39	21	C_1	0.81	0.96	0.88
524	2124	17	4	8	5	C_2	0.87	0.79	0.83
36	273	2250	52	30	5	C_3	0.85	0.85	0.85
16	7	328	2204	86	5	C_4	0.90	0.83	0.87
26	3	17	163	2116	336	C_5	0.75	0.80	0.77
8	1	2	9	533	2100	C_6	0.85	0.79	0.82

(a) Confusion matrix for an LSTM network trained on acceleration data.

(b) Classification report for an LSTM network trained on acceleration data.

4.2 CNN

We also provide the results of our experiments using CNNs for all four modalities. The modalities with sequential data are passed through the network as one data point with an additional temporal dimension. CNNs are state-of-the-art networks for image classification since they are highly able to recognize patterns in images. CNNs are designed according to multilayer perceptrons to reduce processing requirements and they usually consist of a combination of convolutional layers, pooling layers, fully connected layers, and normalization or reshaping layers. For our experiments, we use a CNN with eight layers: Three convolutional layers, two pooling layers, one reshaping layer, and two fully connected layers. The CNN was trained using the Adam Optimizer and cross-entropy as loss function. Again, we randomly select 20% of our dataset to use as a test set and train the CNN on the remaining 80% in ten epochs.

RGB. After training the CNN on the training set of the RGB modality, we achieve an average gesture classification accuracy on the test set of 85.45%. Table 6a shows the confusion matrix for the unimodal gesture classification of RGB data, while Table 6b shows the precision, recall and F1-score for every class.

Table 6. Results for the unimodal classification on the RGB modality using a CNN.

Predicted class $C_{[1-6]}$					
2660	24	1	0	2	1
0	2412	21	15	39	195
0	138	2464	19	23	2
0	55	79	2441	71	0
0	202	2	117	1611	729
0	271	0	45	274	2063

(Target $C_{[1-6]}$)

Class	Precision	Recall	F1-Score
C_1	1.00	0.99	0.99
C_2	0.78	0.90	0.83
C_3	0.96	0.93	0.95
C_4	0.93	0.92	0.92
C_5	0.80	0.61	0.69
C_6	0.69	0.78	0.73

(a) Confusion matrix for a CNN trained on RGB data.

(b) Classification report for a CNN trained on RGB data.

3D. After training the CNN on the training set of the 3D modality, we achieve an average gesture classification accuracy on the test set of 94.05%. Table 7a shows the confusion matrix for the unimodal gesture classification of RGB data, while Table 7b shows the precision, recall and F1-score for every class.

Table 7. Results for the unimodal classification on the 3D modality using a CNN.

Predicted class $C_{[1-6]}$					
2687	1	0	0	0	0
3	2666	1	10	1	1
1	7	2619	17	2	0
0	0	20	2620	1	0
3	1	0	2	2376	279
0	2	4	0	589	2058

(Target $C_{[1-6]}$)

Class	Precision	Recall	F1-Score
C_1	0.94	1.00	1.00
C_2	1.00	0.99	0.99
C_3	1.00	0.99	0.99
C_4	0.99	0.99	0.99
C_5	0.80	0.89	0.84
C_6	0.88	0.78	0.82

(a) Confusion matrix for a CNN trained on 3D data.

(b) Classification report for a CNN trained on 3D data.

Audio. After training the CNN on the training set of the audio modality, we achieve an average gesture classification accuracy on the test set of 45%. Table 8a shows the confusion matrix for the unimodal gesture classification of audio data, while Table 8b shows the precision, recall and F1-score for every class. As can be seen, there is a high recall for the two gestures depending on sound (C_5 and C_6, Snap Once and Twice respectively) while there was a very low recall for the four gestures not depending on sound (C_1 to C_4, Thumbs Up and Down, Swipe Left and Right respectively). As explained in Sect. 3.1, the purpose of the audio modality lies in reinforcing predictions in combination with other modalities [25].

Table 8. Results for the unimodal classification on the audio modality using a CNN (Source: [25]).

Predicted class $C_{[1-6]}$

525	171	207	1771	9	5
460	230	217	1772	2	1
462	168	264	1750	2	0
480	188	178	1798	0	2
85	53	41	60	2076	346
9	6	10	4	381	2243

Target $C_{[1-6]}$

(a) Confusion matrix for a CNN trained on audio data.

Class	Precision	Recall	F1-Score
C_1	0.26	0.19	0.22
C_2	0.28	0.09	0.13
C_3	0.29	0.10	0.15
C_4	0.25	0.68	0.37
C_5	0.84	0.78	0.81
C_6	0.86	0.84	0.85

(b) Classification report for a CNN trained on audio data.

Acceleration. After training the CNN on the training set of the acceleration modality, we achieve an average gesture classification accuracy on the test set of 69%. Table 9a shows the confusion matrix for the unimodal gesture classification of the acceleration data, while Table 9b shows the precision, recall, and F1-scores for every class. As can be seen, it is difficult for the CNN to distinguish Snap Once and Snap Twice (C_5 and C_6 respectively) using only the acceleration modality.

Table 9. Results for the unimodal classification on the acceleration modality using a CNN.

Predicted class $C_{[1-6]}$

2422	5	29	150	53	29
606	1960	37	56	17	6
114	290	1575	616	44	7
109	2	350	2109	65	11
245	2	43	487	1520	364
129	3	15	138	967	1401

Target $C_{[1-6]}$

(a) Confusion matrix for a CNN trained on acceleration data.

Class	Precision	Recall	F1-Score
C_1	0.67	0.90	0.77
C_2	0.86	0.73	0.79
C_3	0.77	0.60	0.67
C_4	0.59	0.80	0.68
C_5	0.57	0.57	0.57
C_6	0.57	0.53	0.63

(b) Classification report for a CNN trained on acceleration data.

5   Multi-modal Fusion

The results of our experiments on unimodal prediction show that both the visual modalities and the acceleration modality have difficulties to distinguish the two classes with little movement but distinct audio (C_5 and C_6 respectively). They also show that the audio modality cannot distinguish the four classes with no distinct sound (C_1 to C_4) but leads to acceptable results in the other two classes. Therefore, we investigate the effect of fusing different modalities to increase the prediction accuracy that can be achieved.

Since the sensors output the data in different formats and also in different frequencies, we discard early fusion methods that fuse the data before using them as input to the machine learning model. We choose two commonly used and easy-to-implement late fusion strategies to prove our assumption, other late fusion strategies as well as intermediate fusion strategies are possible and might even lead to better results but are not within the scope of this work.

We analyse the LSTM or CNN readout layer r^m after the whole gesture sample has been processed, where m denotes the modality $m \in \mathcal{M} = \{\text{RGB, 3D, Audio, Acc}\}$. The readout layer provides one entry for every available gesture class, every entry has a value of $r_i^m \in [0,1]$ and all entries are normalized $\sum_i r_i^m - 1 \forall m \in \mathcal{M}$. $\mathcal{C}$ denotes the decision based on the fused modalities, while $\mathcal{C}^m$ denotes the decisions based on the uni-modal predictions. (Cf. [24])

The first late fusion strategy is called **max-conf**. Here, the uni-modal prediction of the modality with the highest confidence is used, as described in Eq. 8.

$$x = \text{argmax}_{m \in \mathcal{M}} \left(\max_i r_i^m \right)$$
$$\mathcal{C} = \mathcal{C}^x \tag{8}$$

The second late fusion strategy is called **prob**. Here, the readout layer entries are treated as independent conditional probability distributions for a class i given the uni-modal input sequence x^m (cf. [24]). We denote the probabilities as $r_i^m = p^m \left(\mathcal{C}^m = i | x^m \right)$ and use the class with the highest probability after multiplying the independent conditional probabilities, as shown in Eq. 9. (Cf. [24])

$$\mathcal{C} = \text{argmax}_i \left(\prod_{m \in \mathcal{M}} p \left(\mathcal{C} = i | x^m \right) \right)$$
$$= \text{argmax}_i \left(\prod_{m \in \mathcal{M}} r_i^m \right) \tag{9}$$

Figure 10 exemplarily shows some of the results of our multi-modal experiments. It can be seen that the results – that are already very high – can be further improved, even

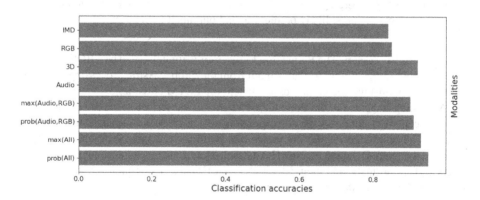

Fig. 10. Gesture classification accuracies achieved by the LSTM networks trained on uni-modal data (Acceleration data, RGB data, 3D data, Audio data) and by performing multi-modal fusion: max-conf of audio and RGB data, prob of audio and RGB as well as max-conf and prob of all four modalities. (Source: [25]).

with rather simple fusion strategies as used here. While fusing two modalities like audio and RGB does not yield better or the same results as uni-modal prediction on 3D data, these results show that fusing two modalities can improve the prediction accuracy, i.e. when a 3D sensor is not available or processing 3D data is not computationally feasible or even possible.

6   Live Demonstrator

Since our choice of only one person performing all the gestures in the dataset, we want to prove that the dataset still can be used to train a real-live system that is able to correctly classify hand gestures that are performed by different people. Thus, experiments on the live system prove that a machine learning model trained on the MMHG dataset is able to generalize to another person apart from the one who recorded all gesture samples.

As a proof of concept, we implement a live system that consists of an LSTM network trained only on the 3D modality of the MMHG dataset and the live classifier based on the Robot Operating Systems (ROS) that receives and processes 3D data from an Orbbec Astra as described in Sect. 3.3 and then feeds it into the trained LSTM network. Afterward, it receives the prediction and outputs it to the user. The other three modalities can be handled accordingly, the network model can be swapped out by another pre-trained model, i.e. a CNN, or fusion can be implemented as well, if needed.

6.1   Implementation

The implementation consists of a Point Cloud Processor that receives the data stream from our 3D sensor and processes it according to the preprocessing steps described in Sect. 3.3. It also consists of an LSTM classifier that receives the preprocessed data from the Point Cloud Processor and feeds it into multiple pre-trained LSTM networks. The third part is the Aggregator which collects the output predictions from the LSTM networks, selects the most likely prediction, and outputs it to the user.

Point Cloud Processor. The Point Cloud Processor is implemented as a ROS node and is responsible for receiving the data stream from the Orbbec Astra, processing the point clouds, and publishing the preprocessed data to the LSTM classifier.

According to Sect. 3.3, the Point Cloud Processor subscribes to the 3D camera sensor. It accepts 3D data at 6 Hz corresponding to the frequency used in the MMHG dataset. The node receives the 3D image and performs the same preprocessing steps performed on the 3D data in the MMHG dataset. Thus, it downsamples the point clouds, infers surface normals, and then calculates Point Feature Histograms. Those PFHs are then published for further processing by the other nodes.

LSTM Classifier. The LSTM Classifier is also implemented as a ROS node and is responsible for performing gesture classification by passing the PFHs published by the Point Cloud Processor through multiple pre-trained LSTM networks. Since gestures can start at any given moment in time, we use an approach called *Shifted Recognizer* [26]: N identical classifiers or recognizers are run simultaneously. Each classifier is pre-trained on the MMHG dataset and has learned to classify gestures with a fixed length $T = 12$ which determines the Temporal Receptive Field (TRF) (accordin to [26]). The LSTM Classifier node feeds the PFHs it receives from the Point Cloud Processor to all N classifiers. Each classifier has a delay of $\Delta = \frac{T}{N}$ frames compared to the other classifiers. Therefore, "if we run enough parallel classifiers, a gesture of length $l \leq T$ will always correlate with the TRF of a single classifier which will then classify it and report its prediction" [25]. Figure 11 shows $N = 4$ parallel classifiers or recognizers denoted as R_n with $n \in [1, N]$. The delay between the classifiers is set to $\Delta = \frac{T}{N}$. A performed gesture will fall into exactly one classifiers TRF (indicated as a green bar in the figure), therefore this classifier will predict the gesture and output the results. All other classifiers (indicated as red bars) will only receive part of the gesture in their TRF, therefore they will not predict the gesture correctly.

In our system, we use $N = 12$ LSTM classifiers since the gesture samples in the MMHG dataset have 12 frames. Therefore, as described above, a gesture will always correlate to exactly one classifier with no onset or offset.

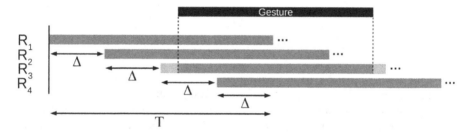

Fig. 11. Shifted Recognizer with $N = 4$. The delay is set to $\Delta = \frac{T}{4}$. The red and green bars indicate the TRF of the four Shifted Recognizers (denoted $R_{1...4}$). The currently performed gesture is shown as a black bar on the top. The current gesture fits in exactly one classifiers TRF, thus it can predict the gesture correctly. (Source: [25]). (Color figure online)

Aggregator. The Aggregator is the third ROS node. It gathers the predictions from the readout layers from all $N = 12$ LSTM classifiers. According to previous research [26] LSTM networks are able to classify sequences with varying onset and offset to some extent. Thus, it is very likely that not just the LSTM classifier in whose TRF the gesture fits in completely but also some of the other LSTM classifiers will predict the correct gesture. Therefore, the Aggregator chooses the gesture class with the highest prediction score, but only if it exceeds a predefined threshold and has been stable for the past three frames. Thus, no prediction is chosen if there is no gesture in the data stream.

6.2 Experiments

We asked four people to perform gestures and had them classified by our live system. Neither of those people were the person who conducted the gestures in the MMHG

dataset. Each person was given a short introduction on how to perform the six different gestures correctly, then they performed gestures from every gesture class and we
recorded the prediction made by the live system.

There were two male and two female users with different hand sizes and skin colors to present as much variation as possible to our system and test the capability of our
system to generalize to different users (cf. [25]). Table 10 shows the results of our experiments. P_i denotes the ratio of correct classifications of a gesture class C_j (as described
in Sect. 3.1) for the i-th user with $i \in \{1, 2, 3, 4\}$.

Table 10. Resulting gesture classification accuracies using our live system trained on the 3D
modality of the MMHG dataset (Source: [25]).

	P_1	P_2	P_3	P_4	Total
C_1	5/5	1/1	1/1	3/3	100%
C_2	5/5	1/1	1/1	2/3	90%
C_3	5/5	1/1	0/1	2/3	80%
C_4	2/5	0/1	1/1	1/3	40%
C_5	5/5	1/1	1/1	2/3	90%
C_6	0/5	0/1	0/1	1/3	10%
$\sum$	73.3%	66.7%	66.7%	61.1%	66.7%

Since our live system is meant as a proof of concept and depends only on the 3D
modality, it is not surprising that the system is not able to distinguish between the gesture classes Snap Once and Snap Twice (C_5 and C_6 respectively). Using multi-modal
fusion of the 3D data with the audio data will most likely improve those results. Also,
for our system Swipe Left and Thumbs Up (C_3 and C_1 respectively) are difficult to
distinguish since the angle and movement of the hand is similar (cf. [25]). Again, using
multi-modal fusion of the 3D data with – for example – acceleration data, could possi-
bly improve the gesture classification accuracies for that gesture class (C_3).

Nevertheless, the experiments on our live system show that a system trained on the
MMHG dataset is able to generalize to other people performing the gestures. The results
can be improved by the use of multi-modal fusion with one or more other modalities.

7   Conclusion

We provide an in-depth description of the new, freely available Multi-Modal Hand Gesture Dataset consisting of almost 80,000 samples in six gesture classes with the four
sensory modalities RGB, 3D, audio, and acceleration. The gesture classes of the dataset
were carefully chosen to be easy to perform by all users and also suitable for application
oriented experiments on sequence classification and multi-modal fusion.

It can be seen that even very simple late-fusion techniques can be combined with
state-of-the-art sequence classification models such as LSTM and CNN models, thus

improving the results of uni-modal gesture classification. Unsurprisingly, the audio modality alone leads to disappoint gesture classification accuracies, but can improve the quality of gesture classification when being fused with other modalities. Of course this stems from the fact that our gesture classes were specifically chosen to show this kind of behavior, to allow the dataset to be well suited for research on multi-modal fusion.

Also, experiments conducted on a live system prove that a generalization to other persons is high even though only a single person recorded all gesture samples available in the dataset.

Future research will include further experiments on generalization capabilities and the possible bias in recognition due to the single subject in the dataset. Also, we will perform experiments using probabilistic models for multi-modal sequence classification, outlier detection, and sampling. Another focus in future work will be more complex – intermediate – fusion strategies, i.e. with an end-to-end learned fusion contribution at multiple stages in a network.

References

1. Angelaki, D.E., Gu, Y., DeAngelis, G.C.: Multisensory integration: psychophysics, neurophysiology, and computation. Curr. Opin. Neurobiol. **19**(4), 452–458 (2009)
2. Beauchamp, M.S.: See me, hear me, touch me: multisensory integration in lateral occipital-temporal cortex. Curr. Opin. Neurobiol. **15**(2), 145–153 (2005)
3. Becker, S., Ackermann, M., Lapuschkin, S., Müller, K.R., Samek, W.: Interpreting and explaining deep neural networks for classification of audio signals (2018)
4. Chen, C., Jafari, R., Kehtarnavaz, N.: Improving human action recognition using fusion of depth camera and inertial sensors. IEEE Trans. Hum.-Mach. Syst. **45**, 51–61 (2014). https://doi.org/10.1109/THMS.2014.2362520
5. Ernst, M.O., Banks, M.S.: Humans integrate visual and haptic information in a statistically optimal fashion. Nature **415**(6870), 429–433 (2002)
6. Escalera, S., et al.: Chalearn multi-modal gesture recognition 2013: grand challenge and workshop summary. In: Proceedings of the 15th ACM on International Conference on Multimodal Interaction, pp. 365–368 (2013)
7. Gepperth, A.R., Hecht, T., Gogate, M.: A generative learning approach to sensor fusion and change detection. Cogn. Comput. **8**(5), 806–817 (2016)
8. Guan, Y., Zheng, M.: Real-time 3D pointing gesture recognition for natural HCI. In: 2008 7th World Congress on Intelligent Control and Automation, pp. 2433–2436 (2008). https://doi.org/10.1109/WCICA.2008.4593304
9. Hochreiter, S., Schmidhuber, J.: Long short-term memory. Neural Comput. **9**, 1735–1780 (1997). https://doi.org/10.1162/neco.1997.9.8.1735
10. Imran, J., Raman, B.: Evaluating fusion of RGB-D and inertial sensors for multimodal human action recognition. J. Ambient. Intell. Humaniz. Comput. **11**(1), 189–208 (2019). https://doi.org/10.1007/s12652-019-01239-9
11. Khaire, P., Kumar, P., Imran, J.: Combining CNN streams of RGB-D and skeletal data for human activity recognition. Pattern Recognit. Lett. **115**, 107–116 (2018)
12. Kim, T.K., Cipolla, R.: Canonical correlation analysis of video volume tensors for action categorization and detection. IEEE Trans. Pattern Anal. Mach. Intell. **31**(8), 1415–1428 (2008)
13. Kopuklu, O., Rong, Y., Rigoll, G.: Talking with your hands: scaling hand gestures and recognition with CNNs. In: Proceedings of the IEEE International Conference on Computer Vision Workshops (2019)

14. Liu, K., Chen, C., Jafari, R., Kehtarnavaz, N.: Fusion of inertial and depth sensor data for robust hand gesture recognition. IEEE Sens. J. **14**(6), 1898–1903 (2014). https://doi.org/10.1109/JSEN.2014.2306094

15. Liu, L., Shao, L.: Learning discriminative representations from RGB-D video data. In: Twenty-Third International Joint Conference on Artificial Intelligence (2013)

16. Marin, G., Dominio, F., Zanuttigh, P.: Hand gesture recognition with jointly calibrated leap motion and depth sensor. Multimed. Tools Appl. **75**(22), 14991–15015 (2015). https://doi.org/10.1007/s11042-015-2451-6

17. McConnell, R.: Method of and apparatus for pattern recognition (1986)

18. Memo, A., Minto, L., Zanuttigh, P.: Exploiting silhouette descriptors and synthetic data for hand gesture recognition. In: Giachetti, A., Biasotti, S., Tarini, M. (eds.) Smart Tools and Apps for Graphics - Eurographics Italian Chapter Conference. The Eurographics Association (2015). https://doi.org/10.2312/stag.20151288

19. Memo, A., Zanuttigh, P.: Head-mounted gesture controlled interface for human-computer interaction. Multimed. Tools Appl. **77**(1), 27–53 (2016). https://doi.org/10.1007/s11042-016-4223-3

20. Nasser, K.: Digital Signal Processing System Design: LabVIEW Based Hybrid Programming (2008)

21. Rusu, R.B., Blodow, N., Marton, Z.C., Beetz, M.: Aligning point cloud views using persistent feature histograms. In: 2008 IEEE/RSJ International Conference on Intelligent Robots and Systems, pp. 3384–3391. IEEE (2008)

22. Sachara, F., Kopinski, T., Gepperth, A., Handmann, U.: Free-hand gesture recognition with 3D-CNNs for in-car infotainment control in real-time. In: 2017 IEEE 20th International Conference on Intelligent Transportation Systems (ITSC), pp. 959–964 (2017). https://doi.org/10.1109/ITSC.2017.8317684

23. Sarkar, A., Gepperth, A., Handmann, U., Kopinski, T.: Dynamic hand gesture recognition for mobile systems using deep LSTM. In: Horain, P., Achard, C., Mallem, M. (eds.) IHCI 2017. LNCS, vol. 10688, pp. 19–31. Springer, Cham (2017). https://doi.org/10.1007/978-3-319-72038-8_3

24. Schak, M., Gepperth, A.: On multi-modal fusion for freehand gesture recognition. In: Farkaš, I., Masulli, P., Wermter, S. (eds.) ICANN 2020. LNCS, vol. 12396, pp. 862–873. Springer, Cham (2020). https://doi.org/10.1007/978-3-030-61609-0_68

25. Schak, M., Gepperth, A.: Gesture recognition on a new multi-modal hand gesture dataset. In: ICPRAM (2022)

26. Schak, M., Gepperth, A.: Robustness of deep LSTM networks in freehand gesture recognition. In: Tetko, I.V., Kůrková, V., Karpov, P., Theis, F. (eds.) ICANN 2019. LNCS, vol. 11729, pp. 330–343. Springer, Cham (2019). https://doi.org/10.1007/978-3-030-30508-6_27

27. Tran, T., et al.: A multi-modal multi-view dataset for human fall analysis and preliminary investigation on modality. In: 2018 24th International Conference on Pattern Recognition (ICPR), pp. 1947–1952 (2018). https://doi.org/10.1109/ICPR.2018.8546308

28. Tran, T.H., et al.: A multi-modal multi-view dataset for human fall analysis and preliminary investigation on modality. In: 2018 24th International Conference on Pattern Recognition (ICPR), pp. 1947–1952 (2018). https://doi.org/10.1109/ICPR.2018.8546308

29. Wan, J., Li, S.Z., Zhao, Y., Zhou, S., Guyon, I., Escalera, S.: ChaLearn looking at people RGB-D isolated and continuous datasets for gesture recognition. In: 2016 IEEE Conference on Computer Vision and Pattern Recognition Workshops (CVPRW), pp. 761–769 (2016). https://doi.org/10.1109/CVPRW.2016.100

30. Wan, J., Zhao, Y., Zhou, S., Guyon, I., Escalera, S., Li, S.Z.: ChaLearn looking at people RGB-D isolated and continuous datasets for gesture recognition. In: Proceedings of the IEEE Conference on Computer Vision and Pattern Recognition Workshops, pp. 56–64 (2016)

31. William T. Freeman, M.R.: Orientation histograms for hand gesture recognition. Technical report TR94-03, MERL - Mitsubishi Electric Research Laboratories, Cambridge, MA 02139 (1994)

32. Zhang, Y., Cao, C., Cheng, J., Lu, H.: EgoGesture: a new dataset and benchmark for egocentric hand gesture recognition. IEEE Trans. Multimed. **20**(5), 1038–1050 (2018). https://doi.org/10.1109/TMM.2018.2808769

Adaptive Sampling for Weighted Log-Rank Survival Trees Boosting

Iulii Vasilev(✉) ⬤, Mikhail Petrovskiy ⬤, and Igor Mashechkin ⬤

Computer Science Department of Lomonosov Moscow State University, MSU, Vorobjovy Gory, Moscow 119899, Russia
iuliivasilev@gmail.com, {michael,mash}@cs.msu.su

Abstract. The field of survival analysis is devoted to predicting the probability and time of the occurrence of an event. The global problem is to predict the event probability over time. It has applications in healthcare, credit scoring, etc. The most widely used method for assessing the covariate impacts on survival is the Cox proportional hazards approach. However, the assumption of non-overlapping survival functions usually does not hold on real data, and the linear dependence on features limits the quality of the method. There are tree-based machine learning methods to solve these problems. Usually, to evaluate the difference between the samples, it used the log-rank test. Obtained survival decision tree models also have strong interpretability, they can evaluate the importance of predictors, but they demonstrate inferior performance in comparison to Cox proportional hazards models.

To overcome these issues, this paper proposes a new boosting of the survival decision tree model that uses adaptive sampling and weighted log-rank split criteria. The model iteratively corrects an error in the ensemble. Each decision tree is trained on a sample, taking into account the weights of observations and subsequently adjusting the probabilities of getting into the next sample. We introduce an experimental comparison of the proposed adaptive boosting method against Cox proportional hazard and widely used survival trees and their ensembles: random forest and gradient boosting. Experiments on healthcare datasets show that our model outperforms the state-of-the-art survival models in terms of the following metrics: the concordance index, the integrated Brier score, and the integrated AUC.

Keywords: Machine learning · Time-to-event analysis · Survival analysis · Random survival forest · Cox proportional hazard · Gradient boosting survival analysis · Adaptive boosting

1 Introduction

Time-to-event analysis originated from the idea of predicting the time until an event occurs. Also, survival analysis tries to measure probability and evaluate the importance of features on survival over the event at some time. The main problem is estimating the event probability for each time moment. These methods are used in medicine, insurance, manufacturing, etc. For example, in manufacturing, survival analysis is used to solve

© Springer Nature Switzerland AG 2023
M. De Marsico et al. (Eds.): ICPRAM 2021/2022, LNCS 13822, pp. 98–115, 2023.
https://doi.org/10.1007/978-3-031-24538-1_5

reliability problems, assuming a system failure is an event. In healthcare, the problem is predicting the event time for a patient with a particular disease. The event is the outcome of the patient's treatment: death, relapse, or recovery.

Missing data is a serious problem. In particular, it is necessary to be able to process observations at an unknown time of the event. Then, if the study ended before the event or the patient disappeared from the study, the event is called censored. Henceforth, we consider only right censoring. Also, we consider the application of survival analysis to medical data.

In the case of patient mortality data, the event is death. Discharge or discontinuation of the observation is considered a censored event. In the problem of mortality analysis, a favorable outcome for the patient is expressed by a longer expected time of occurrence of the event.

In the case of patient admission data, the event is the patient's discharge. Death or termination of the observation is considered a censored event. In the problem of hospitalization analysis, a favorable outcome for the patient is expressed by a shorter expected time of occurrence of the event.

The problem of event probability prediction over time formulates in terms of a survival function:

$$S(t) = P(T > t),$$

where t is the time of observation, and T is the random value of the occurrence of the event.

Also, the T distribution determines based on the hazard function:

$$h(t) = -\frac{\partial}{\partial t} \log S(t)$$

In practice (e.g. disease research), the data contain covariate information (e.g. anamnesis). The challenge is to evaluate the covariation impact on survival function. Let us define X as a random vector of variables and T as a non-negative random variable of time. For an observation with a vector x, we define the probability of the event, which does not occur to a time t, as a conditional survival function:

$$S(t \mid x) = P(T > t \mid X = x).$$

Similarly, the conditional hazard function is:

$$h(t \mid x) = -\frac{\partial}{\partial t} \log S(t \mid x).$$

The state-of-the-art method of survival analysis is Cox proportional hazards (CoxPH) [4]. The method assumes that the shape of the hazard function is the same for all observations, and the differences are determined by the scale coefficient. In particular, the model assumes that the log hazard is a linear function of covariates. Thus, there is a constant relationship between the dependent variable and the regression coefficients:

$$h(t \mid x) = h_0(t) \exp\left(x^T \beta\right),$$

where $h_0(t)$ is the baseline hazard function, x is a vector of covariates, and β is a vector of covariates weights.

To predict the survival function for a specific observation, the survival function $S_0(t)$ is constructed based on the Breslow [19] estimate, and shifts by the weights β:

$$S(t \mid x) = S_0(t)^{\exp(x^T \beta)}.$$

By the way, for the given weights β, we can calculate the risk coefficients (hazard ratio) $\exp(\beta)$, which determine the influence of input features.

However, there are several disadvantages:

- The ratio between the two conditional hazard functions does not change over time.
- The weights of the variables remain constant. In medicine, the impact of risk factors can change (for example, a patient is at greater risk after a surgical operation and has a more steady health condition after recovery).
- A linear combination may not have a basis for a subset of variables.

Finally, for real datasets, there are additional obstacles to the applicability of the existing approaches. Firstly, models do not take into account possible data peculiarities. In particular, early and late events have the same effect on the forecast. Secondly, models cannot work directly with categorical features and missing values, which have widespread in real data.

Our goal is to develop a new approach that eliminates the above-mentioned disadvantages. We suggest tree-based models, which outperform existing tree models and Cox proportional hazards on several benchmark public datasets.

The structure of this paper is as follows. Section 2 contains review and discussion of existing survival analysis algorithms: Survival Tree [17], Random Survival Forest [13], and Gradient Boosting Survival Analysis [7]. We provide a detailed description of the proposed approach in Sect. 3 that is based on iterative ensemble construction with adaptive sampling. Section 4 begins with a discussion of existing survival tree accuracy metrics and public healthcare datasets. It is followed by the experimental results of existing and proposed models. We conclude with Sect. 5 with a summary of our contributions, and further research.

2   Related Work

Features in survival analysis are as follows: input variables X to the beginning of the observation, the time T of the event occurrence, and the binary flag E of the event occurrence (henceforth, we consider that the observations with $E = 0$ are censored).

In survival analysis, we need to train a model on available data with the input features X and the target features T and E and predict the survival and hazard functions.

Further, in this section, we briefly review the most used models in survival analysis: the tree-based model Survival Tree and their ensembles: Random Survival Forest, and Gradient Boosting Survival Analysis.

2.1   Survival Tree

The paper [17] presents a tree algorithm based on the idea of recursively splitting a sample into groups with different survival functions. The tree is built starting from the root node and considers all possible intermediate values for each feature from X. Each value generates two partition branches, and the criterion value is calculated through the targets T and E. The best partition is associated with the maximum difference value between partitioned samples.

The root node is split into two children nodes by the log-rank [18] criterion that is the most commonly used to measure differences between survival functions of two groups. A larger value of the log-rank statistic determines a larger difference between the survival functions of the two samples. The log-rank test has optimal power for detecting a difference in samples in the case of proportional hazard functions. The null hypothesis of the test assumes that there is no difference in the survival rate of the two samples.

Suppose that the times of event occurrence are ordered: $\tau_1 < \tau_2 < ... < \tau_K$. Denote numbers of patients in two groups by n_1 and n_2. For these groups, define $N_{1,j}$ and $N_{2,j}$ as numbers of patients at the moment τ_j, and $O_{1,j}$ and $O_{2,j}$ as numbers of events at the moment τ_j. Then, the total number of patients and events at the moment τ_j is $N_j = N_{1,j} + N_{2,j}$ and $O_j = O_{1,j} + O_{2,j}$.

Let us define the expected event number at the moment τ_j as $E_{i,j} = \frac{N_{i,j}O_j}{N_j}, i = 1, 2$. Based on the available data, we obtain statistics of the log-rank criterion:

$$LR = \frac{\sum_{j=1}^{K} w_j \left(O_{1,j} - E_{1,j}\right)}{\sqrt{\sum_{j=1}^{K} w_j^2 E_{1,j} \left(\frac{N_j-O_j}{N_j}\right) \left(\frac{N_j-N_{1,j}}{N_j-1}\right)}}, \tag{1}$$

where $w_j = 1$. The weights show the sensitivity to the time of events.

According to feature vector x, the survival function is defined as the Kaplan-Meier [14] estimate, which builds on a leaf sample for x.

The tree growth is controlled by: the depth of the tree, the number of splitting features, the number of leaves, and the size of nodes.

The main advantage of the method is interpretability in the form of human-understandable rules. For each leaf, there is a set of rules obtained from the root of the leaf. Thus, for a tree with a small depth, an expert can analyze the rules for consistency and correctness.

However, the method has significant disadvantages: tree construction is based on filled data, any decision tree tends to overfit, and a large amount of data is needed to achieve high accuracy of the decision tree. In the case of limited data, the decision tree model is usually used as the base «weak» model in ensembles.

2.2   Random Survival Forest

The aggregation of several model forecasts improves the quality of forecasting and prevents overfitting.

In [13], it defines the Random Survival Forest (RSF) model as a survival trees ensemble [17] that uses an aggregation of forecasts in the following way:

1. From the source dataset, generate N bootstrap samples.

2. For each sample, build a survival tree. The splitting at each node uses P random features. The best partition is associated with a maximum difference between children nodes.
3. Each tree builds until exhausted.

The ensemble error based on $OOB_i, i = 1...N$, where OOB_i is out-of-back sample of i tree. According to the covariate vector x, the OOB forecast is the average prediction over base models with $x \in OOB_i$.

The predicted survival or hazard functions are the average forecasts of ensemble base models. Averaging allows us to enhance quality and avoid overfitting.

The RSF model is controlled by: the size of the ensemble, the size of the sample, the ratio of features for splitting, and the tree growth control parameters.

2.3 Gradient Boosting Survival Analysis

An alternative ensemble approach is Gradient Boosting [7]. Unlike Random Survival Forest, the Gradient Boosting Survival Analysis (GBSA) algorithm [12] uses an iterative tree learning.

The algorithm has devoted the idea of sequential ensemble construction, that each new model uses the errors of the previous models as a target. The GBSA adds a new model with a weight that minimizes the aggregated loss of the new ensemble.

The purpose of the GBSA is to minimize the ensemble error via a loss function L. In time-to-event analysis, the loss function is usually defined as Cox partial likelihood deviation [4]. Denote the training set by $\{(x_i, y_i)\}_{i=1}^n$, the loss function by L, and the ensemble size by M. Also, we denote the prediction model (boosting ensemble) by F_i, where i describes the ensemble size. The loss function L have two arguments: target value y and prediction value $F(x)$ for features x. The algorithm of GBSA consists of the following steps:

1. Find the α that minimizes the total loss:

$$F_0(x) = \underset{\alpha}{\text{argmin}} \sum_{i=1}^n L(y_i, \alpha)$$

2. For associated model number $m = 1$ to M:
   (a) Calculate pseudo-residuals for each observation i:

$$r_{im} = - \left[\frac{\partial L(y_i, F(x_i))}{\partial F(x_i)} \right]_{F(x)=F_{m-1}(x)}$$

   (b) Build a base model (survival tree) $h_m(x)$ on the training set $\{(x_i, r_{im})\}_{i=1}^n$
   (c) Compute the model weight $v_m(0 < v_m < 1)$ based on following equation:

$$v_m = \underset{v}{\text{argmin}} \sum_{i=1}^n L(y_i, F_{m-1}(x_i) + v \cdot h_m(x_i))$$

   (d) Add the base model with associated weight to the ensemble:

$$F_m(x) = F_{m-1}(x) + v_m \cdot h_m(x),$$

3. Define the resulting ensemble as F_M.

For input observation, the prediction function is a weighted sum over ensemble predictions for each time.

The model is controlled by: the loss function, the ensemble size, the weights calculation approach, and the tree growth control parameters.

As well as the random forest method (in Sect. 2.2), the interpretability of the forecast is lost. However, the ensemble methods have a high forecasting performance and resistance to overfitting. It is important to note that both methods (GBSA and RSF) are the ensemble of survival trees (in Sect. 2.1), which uses only filled data. On the other hand, the bagging method (Sect. 3.2) is the ensemble of survival trees with weighted log-rank criteria (Sect. 3.1), which can use continuous, categorical features and missing values.

3 Proposed Approach

The Survival Tree (Sect. 2.1), RSF (Sect. 2.2), and GBSA (Sect. 2.3) approaches use the log-rank criterion for finding a best split. The log-rank assumes uncorrelated censoring and prediction. Also, early and late events have the same impact on survival differences. Finally, these approaches allow only complete data and can not work with categorical features.

In [3, 18], authors notice poor sensitivity of log-rank to a real dataset with early events and suggest using weighted log-rank splitting criteria.

The weighted log-rank statistics use the following weights w_j definition in (1).

In Wilcoxon statistic [8], it uses the events weights as a count of observations N_j at the moment of time τ_j. The early events have greater weights and pay a greater impact on statistical value. Nonetheless, Wilcoxon criterion depends on the censoring groups.

In Peto-Peto criterion [21], it uses the events weights as a survival function value $\hat{S}(\tau_j)$ at the moment of time τ_j. Thus, it is suited for disproportional hazard function cases. However, differences in the censoring groups are not taken into account by the criterion.

In Tarone-Ware criterion [25], it uses the events weights as a square root of a number of observations $\sqrt{N_j}$ at the moment τ_j. It determines greater weights to earlier events, same as in Wilcoxon criterion. In [16], it notes that Tarone-Ware is the «golden mean» among the weighted criteria.

In the previous work [26], we developed decision trees (Sect. 3.1) and bagging (Sect. 3.2) ensemble that take into account missing values and use weighted log-rank criteria. However, at the moment, we have not found studies about the applicability of adaptive boosting ensembles to survival analysis. Based on the developed survival trees, the adaptive boosting model allows to save advantages and enhances the quality. In this paper, we propose further development of a tree models ensemble with weighted log-rank criteria considering them as a boosting model with adaptive sampling.

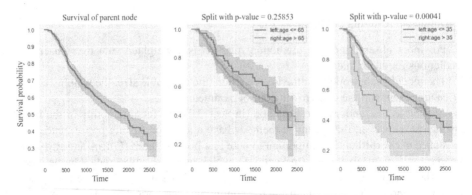

Fig. 1. An example of a parent node splitting by values 65 and 35. For each split, the plot images survival functions and a corresponding p-value. The age value of 35 determines the split with a smaller p-value for child nodes.

3.1   Survival Tree with Weighted Log-Rank Criteria

In [26], we present the new tree-like method, which uses weighted criteria and increases the sensitivity of trained models to data peculiarities. The method uses an algorithm for constructing a binary partition at a decision tree node.

Initially, we choose the best split for each available feature in the node. For continuous feature, intermediate points $a_1, a_2, ...a_k$ is defined as $v_1 < a_1 < v_2 < a_2...a_{n-1} < v_n$, where v_i is a ordered unique values. For each split point a_i, $left$ determined with $v \le a_i$ and $right$ determined with $v > a_i$. Also, we used quantile discretization of points for complexity control. For categorical features, we count all non-overlapping sets l, r of unique values. Then, we obtain $left$ (values from l) and $right$ (values from r) sets of observations.

Then, we count statistic differences (as p-value) for $left$ and $right$ samples. Also, for each sample, in turn, we add missing values and count statistic differences. Finally, the missing values were assigned to the sample with a minimal p-value.

Later, for each available feature, we take splits $left, right$ with a minimum p-value and apply Bonferroni adjustment [1]. The adjustment reduces the significance of the widespread features and gives preference to the rare splits. After all, we choose the feature and its partition with a minimal p-value. The algorithm is repeated recursively for each tree node.

Figure 1 shows a node splitting process by «age» with 65 and 35 values. There are survival functions for the main and child nodes. For a split with age 65, the p-value is equal to 0.25853, and the plots are close to each other. For split with age 35, the p-value is equal to 0.00041, and the right sample (with age over 35) is below the left one. Hence, we take a split with age 35 because it minimized a p-value for child nodes.

The predicted survival function is the Kaplan-Meier estimate for the leaf corresponding to the input observation.

3.2   Survival Bagging with Weighted Log-Rank Criteria

In [26], we present the bagging ensemble of survival decision trees. The bagging ensemble is constructed from a set of «base» independent models with subsequent aggregation of forecasts. The main goal of the «base» model is to describe the training sample accurately. Therefore, approaches to combat overfitting are not applied to the «base» models.

The «base» model is a decision tree Sect. 3.1, built without restriction on the level of significance of partitions and without pruning on the validation set.

The bagging algorithm is based on an iterative decision tree construction. It is made with a bootstrap sample with the same probability for observations. The algorithm continues while the new ensemble model reduces the out-of-back error of the ensemble.

The model forecast is built on the average of ensemble base forecasts. The survival and hazard functions are the mean for each time. The bagging model is controlled by: the size of the ensemble, the size of samples, the tolerance mode indicator, the approach of forecasts aggregation, the type of ensemble error, and the parameters of tree growth control.

3.3   Weighted Boosting

In the Random Forest and Gradient Boosting algorithms, the probabilities of getting observations into the bootstrap subsample are the same. An alternative approach for constructing a boosting ensemble of decision trees is the adaptive boosting algorithm (AdaBoost) proposed by Freund and Schapire [6]. In AdaBoost, each base model is fitted on a sample with weighted observations. Weights allow for an increase or decrease in the significance of the prediction error. After a base model fitting, the observation weights are adjusted, taking into account the base model error. For observations with a low-quality forecast, the weights increase. Using the normalized weights, the next decision tree fits on a more «significant» target of observations.

In [5], Drucker proposes a modification of the AdaBoost algorithm. Unlike AdaBoost, the method uses the weighting of probabilities of getting observations into training subsamples (adaptive sampling). In this case, the next decision tree fits on a more «difficult» subset of observations.

Initially, there are weights $w_i = 1$ for $i = 1, ..., N_1$ for all observations of the training set (with size N_1). The modification has the following steps that are executed until the average loss $\bar{L}$ is less than 0.5 or the number of models in the ensemble is less than M:

1. For an observation i, the probability of getting into the subsample is $p_i = \frac{w_i}{\sum w_i}$. A bootstrap subsample is constructed from N_1 observations.
2. Let t be a decision tree that fits on the bootstrap subsample.
3. Get forecast $y_i^{(t)}(x_i)$ for each observation $i, i = 1, ..., N_1$ of the train sample.
4. Count loss L for each observation: $L_i = L\left(y_i^{(t)}(x_i) - y_i\right)$. The loss L usually has linear (2), square (3), or exponential (4) from:

$$L_i^l = \frac{\left|y_i^{(t)}(x_i) - y_i\right|}{D},$$

(2)

$$L_i^s = \frac{\left|y_i^{(t)}(x_i) - y_i\right|^2}{D^2}, \tag{3}$$

$$L_i^e = 1 - \exp\left[\frac{-\left|y_i^{(t)}(x_i) - y_i\right|}{D}\right], \tag{4}$$

where $D = D = \sup\left|y_i^{(t)}(x_i) - y_i\right|$.

5. Count mean loss $\bar{L} = \sum_{i=1}^{N_1} L_i^f p_i$, where $f \in \{l, e, s\}$.
6. Count measure of confidence $\beta_t = \frac{\bar{L}}{1-\bar{L}}$. Less value of β_t determines high confidence of forecast for tree t.
7. Update the weights of the observation in the training set $w_i \to w_i \beta_t^{(1-L_i^f)}$, where $f \in \{l, e, s\}$. The small loss for the observation leads to a decrease in the weight and probability of the observation getting into the next training sample.

The forecast for the observation with the feature vector x is the weighted sum of the base models t forecasts with normalized weights $\log \frac{1}{\beta_t}$.

$$y(x) = \frac{\sum_{t=1}^{M} \log \frac{1}{\beta_t} y^{(t)}(x)}{\sum_t \log \frac{1}{\beta_t}} \tag{5}$$

The following parameters control the modification of adaptive boosting: the loss function L, the size of ensemble M, and the tree growth control parameters.

3.4 Proposed Adaptive Boosting Ensemble

In this paper, we propose a boosting algorithm based on the iterative construction of a survival trees ensemble. As in [5], we consider a scheme for iteratively updating the probabilities of observations getting into the bootstrap sample:

1. Generate a bootstrap sample with a certain size (specified as a hyperparameter) such that the probability of observations depends on its weight (initialize as 1). Denote by OOB the observations out of the sample.
2. Construct the decision tree on the bootstrap sample (Sect. 3.1).
3. Update the weights of the observation in the training set like AdaBoost does. Each normalized weight is the probability of observations getting into the next bootstrap sample. We use the integrated Brier score [9] as the loss metric. It evaluates the calibration error between forecasting and true survival function.
4. Calculate the OOB-error for the updated ensemble. In terms of OOB-error and an observation x, the prediction is the forecast aggregation over the trees, which contain x in OOB_i.
5. If the new model reduces the OOB-error, then the algorithm goes to step 1. Otherwise, the model eliminates, and the construction breaks.

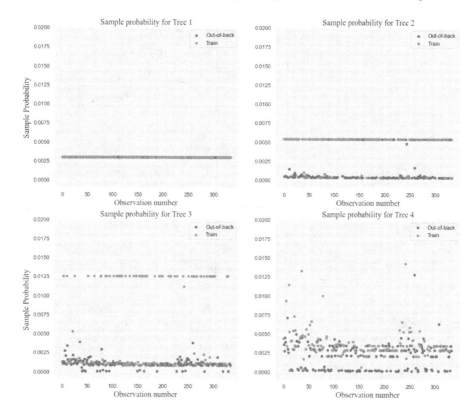

Fig. 2. An example of the probability sample updating for an adaptive boosting model with size = 4. Axis x describes the number of observations in the training set, and axis y describes the corresponding probability. The figures show the probabilities by observation number and final sampling for the new survival tree. For each figure, we highlight sample selection with two colors: orange for the training sample and blue for the out-of-back sample. (Color figure online)

Also, there is tolerance mode: an ensemble constructs for full size N, and the final size figure out by the minimal OOB-error over all iterations.

Figure 2 shows an example of probability updating for the adaptive boosting based on 4 trees. In each figure, we highlight two colors for sample selection: orange for the training sample and blue for the out-of-back sample. The left top plot shows probability initialization for each observation in the training set and data splitting for Tree 1 (into bootstrap and out-of-back samples). The right top plot shows probability after updating weights with Tree 1 and data splitting for Tree 2. Similarly, the left bottom plot and right bottom plot correspond to Trees 3 and 4.

The forecast of the boosting model calculates by Formula (5). The survival and hazard functions calculate as the weighted sum for each time point.

Figures 3 and 4 show an example of prediction for an adaptive boosting model with ensemble size $M = 3$. Figure 3 shows the predicted survival functions of each tree in the ensemble (Tree 1, Tree 2, Tree 3). Figure 4 shows an example of obtaining the final forecast.

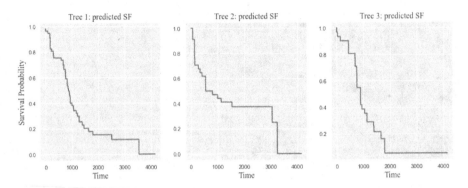

Fig. 3. An example of prediction survival functions for an adaptive boosting model (ensemble size $M = 3$).

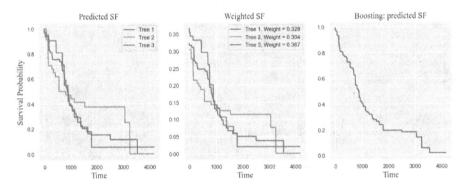

Fig. 4. An example of the final forecast for adaptive boosting. The left plot shows the predicted survival functions for each model. The center plot shows the weighted forecasts corresponding to the tree weight $\log \frac{1}{\beta_t}$. The right plot shows the final survival function (the weighted sum of the predictions of the ensemble models).

To control the computational complexity of the method, we use the following parameters: the ensemble size, the sample size, the tolerance mode flag, the ensemble aggregation metric, the ensemble loss, the observation weight calculation metric, and the tree growth control parameters.

4   Experiments

4.1   Metrics

In this paper, we use following metrics for evaluation the models quality: concordance index [10], integrated Brier score [9] and integrated AUC [11].

The concordance index (CI) is the most commonly used metric for survival analysis. The metric measures the ratio of correctly ordered pairs. The best value of the metric is 1 (proper order), the worst value is 0 (reverse order), and 0.5 reflects the randomness ordering of the model response.

The concordance index calculates as follows:

$$CI = \frac{\sum_{i,j} I(T_j < T_i) \cdot I(\eta_j < \eta_i)}{\sum_{i,j} I(T_j < T_i)},$$

where T_k is the true time, η_k is the predicted time, and I is a binary indicator ($I(T_j < T_i) = 1$ if $T_j < T_i$ else 0).

However, the CI is based only on the event time and can not evaluate the survival function overall. Thus, the metric value is the same for shifted survival functions, although the forecasted and true values can be significantly different.

To overcome these disadvantages, in this paper, we use the integrated Brier score (IBS) [2,9,20], which evaluates the difference between predicted and true survival functions. The true function equals 1 before the event and 0 otherwise.

The Brier score (BS) metric [2] estimates the forecast performance at the fixed time moment t and can be calculated as follows:

$$BS(t) = \frac{1}{N} \sum_i (I(T_i > t) - S(t, x_i))^2 \tag{6}$$

where $S(t, x_i)$ is the forecast at time t for observation x_i with the corresponding event time T_i, and the I is the binary indicator.

If the censorship of observations does not take into account, the following statements are true for a fixed moment t and an observation x_i:

- If the event happened before time t, a low probability of survival (close to 0) is expected,
- If the event occurred after the moment t, a high probability of survival (close to 1) is expected.

Finally, the deviation squares average over all-time moments. The best BS value is 0 when the prediction and truth coincide.

However, the BS metric (6) does not use censoring information. In this case, the Brier score can be modified [9,20] as follows:

$$BS'(t) = \frac{1}{N} \sum_i \begin{cases} \frac{(0 - S(t,x_i))^2}{G(T_i)} & \text{if } T_i \le t, \delta_i = 1 \\ \frac{(1 - S(t,x_i))^2}{G(t)} & \text{if } T_i > t \\ 0 & \text{if } T_i = t, \delta_i = 0 \end{cases} \tag{7}$$

As in (6), $S(t, x_i)$ is a forecast of the survival function at time t for observation x_i with event time T_i. The parameter δ_i (7) is the occurrence indicator of observation x_i. In addition, the $G(t)$ is the Kaplan-Meier estimation, which uses an inverted occurrence flag. Later, the deviation squares (7) need to adjust by the following weights: $\frac{1}{G(T_i)}$ if $T_i \le t$, and $\frac{1}{G(t)}$ if $T_i > t$. Censored observations ahead of time t do not take into account.

Integrated Brier score is used for aggregating the BS estimates over time:

$$IBS = \frac{1}{t_{max}} \int_0^{t_{max}} BS(t) dt$$

An alternative metric for assessing the quality of a forecast is the integrated AUC (IAUC) proposed by Heagerty and Zheng in [11]. They present a method for calculating the ROC curve and area under the curve (AUC) in multi-class or temporary cases. For a fixed time t, divide the observations into two sets that have occurred and non-occurred events. The $\widehat{AUC}(t)$ metric measures the weighted proportion of pairs of observations from each set that have a concerted hazard order (earlier events should have a higher hazard at time t):

$$\widehat{AUC}(t) = \frac{\sum_{i=1}^{n} \sum_{j=1}^{n} I(y_j > t) I(y_i \le t) w_i I(\hat{h}(t, x_j) \le \hat{h}(t, x_i))}{(\sum_{j=1}^{n} I(y_j > t))(\sum_{j=1}^{n} I(y_i \le t) w_i)},$$

where $\hat{h}(t, x_i)$ is the cumulative hazard estimate for observation x_i at time t, w_i is the inverse censoring probability for observation x_i obtained from the Kaplan-Meier estimate ($w_i = G(t_i)$).

Integrated AUC is aggregation $\widehat{AUC}(t)$ score over time:

$$IAUC(t_{min}, t_{max}) = \frac{1}{\hat{S}(t_{min}) - \hat{S}(t_{max})} \int_{t_{min}}^{t_{max}} \widehat{AUC}(t) d\hat{S}(t)$$

Thus, to evaluate the forecasting performance, this paper considers metrics:

1. Concordance index evaluates the correctness of the event time ordering,
2. Integrated AUC evaluates the correctness of the predictions ordering for the cumulative hazard function,
3. Integrated Brier score evaluates the calibration error for survival function.

4.2 Datasets

In this paper, we consider the following medical datasets:

1. PBC [15] – dataset from the Mayo Clinic (diagnosis K74.3)
2. GBSG [24] – dataset from the German Breast Cancer Study (diagnosis C50)
3. Wuhan [27] – dataset of patients with COVID-19 from Wuhan, China.

The dataset on Primary Biliary Cirrhosis (PBC) was collected from 1974 to 1984. A lethal outcome is considered an event. There are 276 observations and 17 features, which determine the status of cirrhosis, optimal treatment strategy, and clinical and laboratory tests. Also, the PBC contains five categorical features: trt, sex, ascites, hepato, spiders. The twelve features of PBC (including treatment strategies and clinical indicators) contain missing. The maximal missing numbers are contained in the cholesterol (134 missing values) and the triglyceride (136 missing values) tests. By the end of the follow-up, 263 patients were without lethal outcomes.

The dataset from the German Breast Cancer Study Group (GBSG) was collected from 1984 to 1989. The relapse of breast cancer is considered an event. There are 686 observations and 8 features, which determine the characteristics of the tumor and optimal treatment strategies. The GBSG dataset contains three categorical features: htreat

Table 1. Hyperparameters for models.

Predictive model	Hyperparameter	Values
CoxPH survival analysis	Regularization penalty	0.1, 0.01, 0.001
	Ties	Breslow, efron
Survival tree	Split strategy	Best, random
	Max depth	From 10 to 30 step 5
	Min leaf size	From 1 to 20 step 2
	Max features	Sqrt, log2, None
Random survival forest	Num estimators	From 10 to 100 step 10
	Max depth	From 10 to 30 step 5
	Min leaf size	From 1 to 20 step 2
	Max features	Sqrt, log2, none
Gradient boosting SA	Num estimators	From 10 to 100 step 10
	Max depth	From 10 to 30 step 5
	Min leaf size	From 1 to 20 step 2
	Max features	Sqrt, log2, none
	Loss function	Coxph, squared, ipcwls
	Learning rate	From 0.01 to 0.5 step 0.01
Tree	Max depth	From 10 to 30 step 5
	Min leaf size	From 1 to 20 step 1
	Significance threshold	0.01, 0.05, 0.1, 0.15
Bagging	Bootstrap sample size	From 0.3 to 0.9 step 0.1
	Num estimators	From 10 to 100 step 5
	Max depth	From 10 to 30 step 2
	Min leaf size	From 1 to 20 step 1
Boosting	Bootstrap sample size	From 0.3 to 0.9 step 0.1
	Num estimators	From 10 to 100 step 5
	Max depth	From 10 to 30 step 2
	Min leaf size	From 1 to 20 step 2
	Ensemble metric	ibs, c_index
	Weighted sum	True, false

(hormonal therapy), menostat (menopausal status), and tumgrad (tumor grade). There are no missings in the GBSG dataset. By the end of the follow-up, 387 patients were without relapse.

The Wuhan dataset was collected from January 10 to February 18, 2020, and presented in [27]. The time of the patient's discharge is considered an event. There are 375 observations and 76 features, which determine an anamnesis and clinical findings during treatment. The feature space builds from the minimum, maximum, and average scores of the patient clinical tests. All features can contain missings, and the maximum missing number in antithrombin and fibrin breakdown products (173 missing values). By the end of the follow-up, 174 patients were without discharge.

4.3   Experimental Setup

Initially, we preprocess the datasets and form the feature space and target variables (time before the event, censoring flag). Then, we perform cross-validation [23] with 5 folds

Table 2. GBSG dataset results.

Predictive model	CI	IBS	IAUC
CoxPHSurvivalAnalysis	0.61544	0.17628	0.71584
SurvivalTree	0.58911	0.19151	0.64322
RandomSurvivalForest	0.61777	0.17504	0.72281
GradientBoostingSurvivalAnalysis	0.61627	0.19331	0.71959
Tree (Peto-Peto)	0.59089	0.19397	0.68932
Tree (Tarone-Ware)	0.59915	0.18586	0.68986
Tree (Wilcoxon)	0.59887	0.18331	0.69028
Tree (Log-rank)	0.58786	0.18934	0.67687
BootstrapTree (Peto-Peto)	0.61871	0.17236	0.73048
BootstrapTree (Tarone-Ware)	0.62162	0.17418	0.73173
BootstrapTree (Wilcoxon)	0.62281	0.17137	0.73456
BootstrapTree (Log-rank)	0.62331	0.17102	0.73165
BoostingTree (Peto-Peto)	**0.63099**	0.17103	0.73366
BoostingTree (Tarone-Ware)	0.62298	0.17164	0.73383
BoostingTree (Wilcoxon)	0.61875	0.17622	0.73303
BoostingTree (Log-rank)	0.62557	**0.16958**	**0.73924**

on the grid of hyperparameters (selected hyperparameters are presented in the Table 1). Cross-validation involves dividing the original sample into five non-overlapping parts, where four are used to train the model (the training set contains 80% of the observations), and one part is used to test the model and calculate quality metrics. In this case, there are five iterations of training/testing the model, where each part is used once as a test sample.

At the time of testing the model, we calculate the following quality metrics: CI (estimates the predicted time of the event), $IAUC$ (estimates the predicted cumulative hazard function), and IBS (estimates the predicted survival function). The resulting cross-validation metric is the average value of the metric over all iterations.

According to the cross-validation results, the calculated metrics correspond to the selected hyperparameters. The best hyperparameters for the model are determined by the minimum IBS over cross-validation.

4.4   Results

For the existing approaches (CoxPH, Survival Tree, RSF, GBSA), we used the scikit-survival library [22] implementation. For the proposed methods (Tree, Bagging, Boosting), we implemented our own code. For each proposed model, we consider four weighted log-rank criteria.

The performance of estimating for all methods are presented in Tables 2, 3, and 4 (the best result by each metric is marked with bold).

For the GBSG dataset (the problem of mortality analysis), the proposed adaptive boosting approach retains two out of three places in terms of the best value for each

Table 3. PBC dataset results.

Predictive model	CI	IBS	IAUC
CoxPHSurvivalAnalysis	0.63886	0.23128	0.7486
SurvivalTree	0.62985	0.29417	0.71021
RandomSurvivalForest	0.66176	0.2138	0.79107
GradientBoostingSurvivalAnalysis	0.65848	0.24946	0.78322
Tree (Peto-Peto)	0.62684	0.25655	0.74223
Tree (Tarone-Ware)	0.64873	0.27711	0.74629
Tree (Wilcoxon)	0.63631	0.24615	0.74749
Tree (Log-rank)	0.63353	0.23548	0.76233
BootstrapTree (Peto-Peto)	**0.66665**	0.19739	0.79075
BootstrapTree (Tarone-Ware)	0.65899	0.20317	0.79243
BootstrapTree (Wilcoxon)	0.66096	0.21042	0.79524
BootstrapTree (Log-rank)	0.66399	**0.19226**	0.79677
BoostingTree (Peto-Peto)	0.66232	0.20656	0.79744
BoostingTree (Tarone-Ware)	0.66448	0.21184	0.79782
BoostingTree (Wilcoxon)	0.65611	0.20723	0.78512
BoostingTree (Log-rank)	0.65568	0.20276	**0.80031**

Table 4. Wuhan dataset results.

Predictive model	CI	IBS	IAUC
CoxPHSurvivalAnalysis	0.70672	0.17098	0.66714
SurvivalTree	0.66951	0.16176	0.67725
RandomSurvivalForest	0.70611	0.13251	0.73522
GradientBoostingSurvivalAnalysis	0.74251	0.13492	0.74467
Tree (Peto-Peto)	0.69466	0.13262	0.73758
Tree (Tarone-Ware)	0.69129	0.13578	0.73296
Tree (Wilcoxon)	0.69913	0.13757	0.7343
Tree (Log-rank)	0.69524	0.14711	0.71774
BootstrapTree (Peto-Peto)	0.75495	0.12432	0.75212
BootstrapTree (Tarone-Ware)	0.75962	0.12497	0.75285
BootstrapTree (Wilcoxon)	0.76266	0.12629	**0.75787**
BootstrapTree (Log-rank)	0.75204	0.12624	0.75085
BoostingTree (Peto-Peto)	0.76593	0.12575	0.74751
BoostingTree (Tarone-Ware)	**0.77369**	**0.11874**	0.74108
BoostingTree (Wilcoxon)	0.7634	0.12344	0.75114
BoostingTree (Log-rank)	0.75735	0.11981	0.75044

metric. According to the totality of metrics, the best method is adaptive boosting with a log-rank criterion. This method shows the best IBS and IAUC metrics. In terms of the CI metric, the best result is the boosting algorithm with the Peto-Peto criterion. Note that all proposed ensembles (bagging and boosting) outperform the existing methods from scikit-survival in all metrics.

For the PBC dataset (the problem of mortality analysis), the proposed adaptive boosting approach retains three places in terms of the best IAUC metric. Based on the set of metrics, boosting methods with log-rank and Tarone-Ware criteria are the best. According to the CI and IBS metrics, the best result has the proposed bagging ensemble with log-rank and Peto-Peto criteria. Note also that all proposed ensembles outperform the existing methods in terms of IBS and IAUC metrics, and in terms of CI metrics, 5 out of 8 proposed ensembles show the best results.

For the Wuhan dataset (the problem of hospitalization analysis), the proposed adaptive boosting approach retains the top three places in terms of CI and IBS metrics. According to the totality of metrics, boosting ensembles with the Tarone-Ware and Wilcoxon criteria are the best. According to the IAUC metric, the best result has the proposed bagging ensemble. Also, all proposed ensembles outperform the existing methods in all metrics.

It is important to note that our implementation of decision trees shows better results (according to the totality of metrics) compared to the existing Survival Tree for all datasets.

5   Conclusions

In this paper, we proposed a method of constructing a nonlinear boosting ensemble of survival trees with adaptive sampling. The idea is to fit the next decision tree on a more «difficult» subset of observations. The boosting ensemble overcomes some disadvantages of the existing methods. Also, it does not assume the proportionality of hazards over time and linear dependences between the hazard logarithm and a feature combination.

The boosting ensemble can handle missing values and categorical features and can pay more attention to early events. The Bonferroni adjustment allows the selection of the best feature with the correct comparison of significant splits.

The experiments included the real medical datasets PBC, GBSG, and Wuhan for two tasks of survival analysis: analysis of mortality and hospitalization of patients. According to the results of the experiments, the proposed boosting algorithm outperforms the existing methods CoxPH, RSF, and GBSA in terms of the following metrics: concordance index (CI), integrated Brier score (IBS), and integrated AUC (IAUC). Also, the proposed method outperforms the previously developed tree bagging ensemble for the mortality analysis problem and shows the best CI and IBS values for the hospitalization problem.

In further research, we are planning to investigate efficient algorithms for selecting optimal split for high-dimensional categorical features. Also, we are planning to explore the time and memory performances of the proposed approaches on real datasets from alternative application areas of survival analysis.

References

1. Benjamini, Y., Hochberg, Y.: Controlling the false discovery rate: a practical and powerful approach to multiple testing. J. Roy. Stat. Soc.: Ser. B (Methodol.) **57**(1), 289–300 (1995)
2. Brier, Glenn W.., Allen, Roger A..: Verification of weather forecasts. In: Malone, Thomas F.. (ed.) Compendium of Meteorology, pp. 841–848. Springer, Boston (1951). https://doi.org/10.1007/978-1-940033-70-9_68

3. Buyske, S., Fagerstrom, R., Ying, Z.: A class of weighted log-rank tests for survival data when the event is rare. J. Am. Stat. Assoc. **95**(449), 249–258 (2000)
4. Cox, D.R.: Regression models and life-tables. J. Roy. Stat. Soc.: Ser. B (Methodol.) **34**(2), 187–202 (1972)
5. Drucker, H.: Improving regressors using boosting techniques. In: ICML, vol. 97, pp. 107–115. Citeseer (1997)
6. Freund, Y., Schapire, R.E., et al.: Experiments with a new boosting algorithm. In: ICML, vol. 96, pp. 148–156. Citeseer (1996)
7. Friedman, J.H.: Greedy function approximation: a gradient boosting machine. Ann. Stat. **29**(5), 1189–1232 (2001)
8. Gehan, E.A.: A generalized Wilcoxon test for comparing arbitrarily singly-censored samples. Biometrika **52**(1–2), 203–224 (1965)
9. Haider, H., Hoehn, B., Davis, S., Greiner, R.: Effective ways to build and evaluate individual survival distributions. J. Mach. Learn. Res. **21**, 85–1 (2020)
10. Harrell, F.E., Jr., Lee, K.L., Mark, D.B.: Multivariable prognostic models: issues in developing models, evaluating assumptions and adequacy, and measuring and reducing errors. Stat. Med. **15**(4), 361–387 (1996)
11. Heagerty, P.J., Zheng, Y.: Survival model predictive accuracy and ROC curves. Biometrics **61**(1), 92–105 (2005)
12. Hothorn, T., Bühlmann, P., Dudoit, S., Molinaro, A., Van Der Laan, M.J.: Survival ensembles. Biostatistics **7**(3), 355–373 (2006)
13. Ishwaran, H., Kogalur, U.B., Blackstone, E.H., Lauer, M.S.: Random survival forests. Ann. Appl. Stat. **2**(3), 841–860 (2008)
14. Kaplan, E.L., Meier, P.: Nonparametric estimation from incomplete observations. J. Am. Stat. Assoc. **53**(282), 457–481 (1958)
15. Kaplan, M.M.: Primary biliary cirrhosis. N. Engl. J. Med. **335**(21), 1570–1580 (1996)
16. Klein, J.P., Moeschberger, M.L.: Statistics for biology and health. Stat. Biol. Health New York **27238**, 205–223 (1997)
17. LeBlanc, M., Crowley, J.: Survival trees by goodness of split. J. Am. Stat. Assoc. **88**(422), 457–467 (1993)
18. Lee, S.H.: Weighted log-rank statistics for accelerated failure time model. Stats **4**(2), 348–358 (2021)
19. Lin, D.: On the Breslow estimator. Lifetime Data Anal. **13**(4), 471–480 (2007)
20. Murphy, A.H.: A new vector partition of the probability score. J. Appl. Meteorol. Climatol. **12**(4), 595–600 (1973)
21. Peto, R., Peto, J.: Asymptotically efficient rank invariant test procedures. J. R. Stat. Soc.: Ser. A (Gener.) **135**(2), 185–198 (1972)
22. Pölsterl, S.: scikit-survival: a library for time-to-event analysis built on top of scikit-learn. J. Mach. Learn. Res. **21**(212), 1–6 (2020)
23. Refaeilzadeh, P., Tang, L., Liu, H.: Cross-validation. Encycl. Database Syst. **5**, 532–538 (2009)
24. Schumacher, M.: Rauschecker for the German breast cancer study group, randomized 2× 2 trial evaluating hormonal treatment and the duration of chemotherapy in node-positive lbreast cancer patients. J. Clin. Oncol. **12**, 2086–2093 (1994)
25. Tarone, R.E., Ware, J.: On distribution-free tests for equality of survival distributions. Biometrika **64**(1), 156–160 (1977)
26. Vasilev, I., Petrovskiy, M., Mashechkin, I.: Survival analysis algorithms based on decision trees with weighted log-rank criteria. In: Proceedings of the 11th International Conference on Pattern Recognition Applications and Methods, pp. 132–140 (2022). https://doi.org/10.5220/0000155500003122
27. Yan, L., et al.: An interpretable mortality prediction model for COVID-19 patients. Nat. Mach. Intell. **2**(5), 283–288 (2020)

Applications

Forecasting Overtime Budgets for Naval Fleet Maintenance Facilities Using Time-Series Analysis During Transient System States

Charith Gunasekara$^{(\boxtimes)}$ (ID), Lise Arseneau (ID), and Cheryl Eisler (ID)

Defence Research and Development Canada, Ottawa, ON K1A 0K2, Canada
charith.gunasekara@forces.gc.ca, {lise.arseneau,
cheryl.eisler}@drdc-rddc.gc.ca

Abstract. In 2019, predictive models were initially developed to attempt to better predict an annual budget for staffing overtime hours within a Royal Canadian Navy (RCN) fleet maintenance facility. The H20.ai open-source framework was used, and models were implemented in the R programming language. Model validation at the time showed the predicted hours were within 5% error rate compared to the actual data. However, when it came to re-apply the process to fiscal year 2020/2021 data, the impact of the COVID-19 pandemic on factors such as the workforce and the logistics supply chain, changed the system dynamics sufficiently that the autoML algorithms had difficulty generating accurate estimates. Therefore, it was decided to examine how times series forecasting methods would predict overtime hours at the fleet maintenance facility. Since historical daily data were readily available, the open-source Prophet model developed by Facebook was used because it can incorporate multiple seasonal patterns, as well as variable holiday effects. The models were tested on fiscal years 2019/2020 and 2020/2021, which showed over 90% accuracy in predicting the total overtime hours. The revised approach in this follow-on study was used to provide financial comptrollers with a prediction for fiscal year 2021/2022.

Keywords: Time series analysis · Predictive analytics · Forecasting · Budget · Overtime hours · Fleet maintenance facilities · Work orders

1 Introduction

1.1 Background

Fleet maintenance facilities operated by the Royal Canadian Navy (RCN) [1] require staff, both civilian and military, to complete repair work for which accurate budget estimates must be derived. The complex and often repetitive nature of the maintenance operations means that the tasks are tracked in detail in enterprise resource management systems over time. Establishing a relationship between the tasks completed and the overtime hours accrued by staff is key to accurate budget predictions.

The RCN operates two repair facilities for its naval platforms, FMF Cape Breton (FMF CB) collocated with Canadian Forces Base (CFB) Esquimalt and FMF Cape

© Springer Nature Switzerland AG 2023
M. De Marsico et al. (Eds.): ICPRAM 2021/2022, LNCS 13822, pp. 119–133, 2023.
https://doi.org/10.1007/978-3-031-24538-1_6

Scott (FMF CS) collocated with CFB Halifax. At these facilities, maintenance tasks are tracked using Work Orders (WOs), which capture the nature of task, the client for delivery, platform, and various details related to the task. FMFs process hundreds of thousands of WOs every year, only a portion of which require overtime (OT) to close out. Overtime hours can be accrued by staff completing maintenance tasks, which are driven by internal or external factors such as the operational demand tempo for ready naval vessels, maintenance or human resource policies, supply chain logistics, or resource limitations. As large quantities of detailed data are tracked within the Defence Resource Management Information System (DRMIS) on WOs, the problem becomes amenable to predictive learning approaches that can take advantage of such stores of information.

1.2 Motivation

Financial comptrollers from FMF CB initially proposed a study that was conducted by Holmes [2] to identify factors associated with accrual of overtime and improve the accuracy of budget predictions at a single FMF. The use of such predictive analytics would allow decision makers to prepare more accurate funding estimates over time – potentially reserving funds for upcoming critical maintenance tasks or saving funds through alternative approaches to task management. Once an end-to-end process is developed, OT hours are but a single use case of predictive analytics that can leverage data from existing enterprise resource management systems to demonstrate the real-world applicability of mining of large data sets.

To facilitate trend analysis, FMF CB supplied seven years of their past OT data, budgets, and WO attributes. Major variables of importance for OT accrual were then identified [2]. The prior approach by Eisler and Holmes [4] utilized Automated Machine Learning (autoML) from H2O.ai [3] that predicted both total OT for a given WO and the month within the Fiscal Year (FY) that the OT was accrued based on the attributes of the WO.

It was found that a fitted logistics function provided an improved estimate for annual cumulative OT accrued as a function of time per fiscal year (FY). Tree-based algorithms were informative as to what WO attributes contribute the most to OT, enabling quantification of relative importance. The use of autoML algorithms improved OT budget estimates for the FMF with a maximum error of 5% observed for fiscal year 2019/2020 (FY19/20), and use of multiple, related datasets (i.e., multiple fiscal years) enabled prediction of multiple variables.

One of the primary assumptions in the prior approach, as described in Sect. 2.1 was that the pool of personnel and resources required to complete WOs would not shrink or grow significantly compared to current levels. This assumption was reasonable in a steady-state system, with no significant changes to supply or demand [4]. However, when looking to re-apply the same process to the next fiscal year (FY20/21) to try to predict the OT accrual, the assumption was no longer valid. As a result of the impact of COVID-19 pandemic on the workforce, in terms of restrictions required to comply with public health orders and measures on base, within the logistics supply chain to provide parts to conduct vessel maintenance, and the ability to deploy fully trained, healthy crews, systemic shocks were observed from March 2020 onwards. Because the system was no longer in steady state operation, the autoML algorithms had difficulty generating an accurate estimate for OT for FY20/21.

Several methods were examined in this revised approach to incorporate in-year accrual data into the autoML models, such as to train on a monthly basis to provide faster feedback response to the system. The results of these trials will be discussed in Sect. 2.1. Rather than continue to develop more complex machine learning models, a new approach was needed. Since the historical OT hours are recorded by date, these data are time series data and time series forecasting methods can be applied; the methodology for which will be covered in Sect. 2.2. Results of the revised approach will be presented in Sect. 3 and conclusions regarding their applicability to the FMFs in Sect. 4.

1.3 Related Work

Supervised machine learning algorithms were applied to a similar Canadian defence application where the variables of importance associated with maintenance task completion times were examined [5]. Clustering techniques were utilized to examine WOs by completion time and task type. This study also provided insights on data cleaning and pre-processing from DRMIS extracts.

For the prior approach described in Sect. 1.1, H2O autoML was used in order to predict both the total OT for a given WO and the month within the FY when the OT is accrued based on attributes of the WO [2]. Open-source tools for automating machine learning pipelines from end-to-end to reduce the time and effort spent on repetitive tasks have been compared in the literature (see [6–9]), and H2O autoML was initially selected [4]. Usually, autoML algorithms predict on a single output variable of interest, but there were two variables to predict in the prior approach: the amount of OT hours (a non-negative numeric value) and the month OT is accrued by WO (categorical). There are relatively few machine learning applications that attempt such a combined variable approach (see [10–15]), which informed the prior approach in [4].

Reapplying the techniques to additional fiscal years with sudden changes in system dynamics revealed the inherent limitation of large datasets required when using autoML algorithms, as the algorithms could not readily adapt to recent changes in the system and could not generate an accurate estimate of OT hours for FY20/21. As a result, it was decided to examine other forecasting techniques applied to the same dataset we used in [4] in order to improve the accuracy of predictions.

Since the OT hours by date are time series data, time series forecasting models can be used to project OT hours to the end of the FY. This would enable analysts to extrapolate on trend and seasonal patterns that had not previously been examined. Many time series forecasting algorithms, related to machine learning and statistics, have been proposed in the literature. A literature review was conducted in 2006 that summarized 25 years of publications on time series forecasting [16]. Over 940 papers were categorized according to forecasting model, e.g., exponential smoothing, autoregressive integrated moving average (ARIMA), dynamic regression, and generalized autoregressive conditional heteroskedasticity.

Time series forecasting has many applications, such as stock prices forecasting, weather forecasting, business planning and resource allocation. Time series forecasting has been used in certain military applications. In [17], exponential smoothing models were applied to voluntary Regular Force attrition data by month and determined that seasonal models were most suitable to provide point predictions for six- and twelve-month attrition volumes. Subsequent work in [18] applied two time series models, the Holt-Winter's exponential smoothing and the Kalman filter, to all Regular Force attrition data by month in order to forecast the attrition volume at the end of the FY. Other military applications where time series models are applied include: forecasting enrollments using ARIMA [19]; forecasting future Marine Corps personnel inventory levels using Holt-Winter's exponential smoothing and the Box-Jenkins ARIMA models [20]; and forecasting military expenditures in India using Box-Jenkins ARIMA model [21].

Advanced forecasting methods such as neural networks and the Prophet model developed by Facebook, Inc. Are described in [22]. Practical issues with forecasting are also presented, which include the challenges when the data are a daily time series involving multiple seasonal patterns. When such a time series is long, which it is in the case of OT hours by date for several FYs, it becomes necessary to use Seasonal and Trend decomposition using Loess (STL), dynamic regression or Prophet that allow for regular seasonality. The Prophet model can also easily incorporate seasonality associated with moving holiday effects (e.g., Thanksgiving does not occur on the same date each year) which can influence when OT hours at the FMFs occur [23]. Several studies using Prophet have been documented in the literature where applications include forecasting wholesale food prices [24], air pollution [25] and traffic forecasting on telecom networks [26].

2 Methodology

2.1 Prior Approach

As described by Eisler and Holmes [4], the autoML pipeline was engaged with a human-in-the-loop during the data pre-processing and feature engineering phases. This enabled two separate models to be generated: the amount of OT hours (a non-negative numeric value) and the month OT accrued by WO (categorical), followed by automatic hyper-parameter optimization. Previously built models were used to cross checked against the final model interpretation, and identify correlations between WO variables, and the predictive analysis output for visualization.

Data Pre-processing. Datasets were initially limited to FY13/14 through FY19/20 information extracted from official enterprise records management systems. To predict for two variables of output, it was necessary to have two separate - although linked - datasets from which to build each model. First, the amount of OT accrued was determined from a dataset with all past WOs from FY13/14-FY18/19 (for initial model development, referred to as Dataset A) [4]. An initial snapshot of open WO near the beginning of FY19/20 was collected for predictive purposes (Dataset A_p). These data detailed information associated work order identification (ID), history, maintenance specifics, and financial coding. The second dataset (Dataset B) was limited to only those WO that accrued OT, which detailed work order identification, OT hours by date, and classification of personnel assigned. Both datasets included numerical, categorical and time-series data.

Data cleaning and augmentation steps were completed manually using Microsoft Excel®. A variety of tools were utilized to transform the raw data into features that better represented the underlying problem. Datasets A and B were then stacked to build a model which used all available FYs as training data. Checks for multicollinearity and correlation of dataset fields were completed [6] to address potential issues in the use of autoML techniques. Simple decision and regression trees provided initial insight by displaying variables of importance for feature construction and extraction.

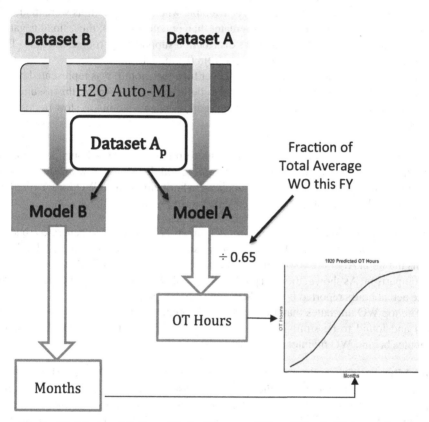

Fig. 1. Models A and B, for predicting Hours per WO and Month OT Accrued [4].

Model Generation. Figure 1 illustrates the two models that were built using H2O autoML; the first predicted OT hours by WO and the second that predicted the month in which the OT hours accrued against the WO. A single model could not predict both OT hours and months because there was insufficient linking data. Dataset A did not contain a date or month column to train the model on, and Dataset B was not suitable to predict OT hours, due to overestimation if OT were accrued for all WO [3].

Four assumptions were made prior to using the predictive outputs of both models [4]:

- The pool of personnel and resources required to complete WOs will not shrink or grow significantly compared to current levels. This assumption is only reasonable in a steady-state system, with no major changes to supply or demand.
- The total number of WO for FY19/20 used to predict on was 65% of the average WO total over the past six years; thus, it is assumed the model predicts 65% of OT for FY19/20, and the actual predicted values must be scaled up accordingly. Over time in a steady state system, as more data becomes available, the estimate should become more accurate. It is important to note that this scaling factor would have to be updated annually, based on the fraction of WO available when the predictions were made.
- OT predicted by the model must be greater than or equal to zero; thus, small negative values predicted from H2O autoML's distribution support were rounded to zero. This was necessary to ensure realistic output from the model.
- The distribution of all WO for FY19/20 over time (by month) was represented by the same distribution of WO currently used to build the model. Again, this assumption is likely only realistic in a steady-state system or when adapting to slow trends over time.

Dataset A was used to build Model A, which predicted the OT hours per WO and Dataset B was used to build Model B, which predicted the Month in which the OT was accrued against the WO. Once these models were built, all planned WO for FY19/20 up to 26 June 2019 (Dataset A_p) were used as input for prediction [2].

Model Interpretation. The previous model output [2] was used to generate a budget prediction graph for FY19/20 by fitting a logistics function for the predicted values[1] results in Fig. 2. After the completion of the fiscal year, the actual OT data was collected for comparison. As shown, the final predicted total of OT was within 5% maximum error of the actual hours reported by the FMF at the end FY19/20.

The top WO attributes that are important for predicting OT accrual were also identified and found to be similar to those of Maybury [5] when analyzing the primary attributes behind WO maintenance hours for the FMFs.

Model Re-use. The model was prepared for use at the beginning of FY20/21 by moving the data from the previously predicted year (Dataset A_p) to Dataset A. A new Dataset A_p was obtained shortly after the new FY began with all WO from 1 Apr to 19 Jun 2020 and the existing model could be re-trained or a new model using the established process could be developed. It was postulated that having this more recent data could also allow for previous years with visibly differentiable trends in reporting to be removed from the input dataset.

However, both the re-trained model and a new model trained on the additional data produced significant overestimations and high variance (~30–50%) of the OT hours required, when compared to the first several months of OT hours accrued and later once additional data became available, due to violation of the steady state assumption of the system.

---

[1] Error bars on the predictive estimates are shown, although they are difficult to discern early on because they are on the order of magnitude of the size of the marker at the start of the FY.

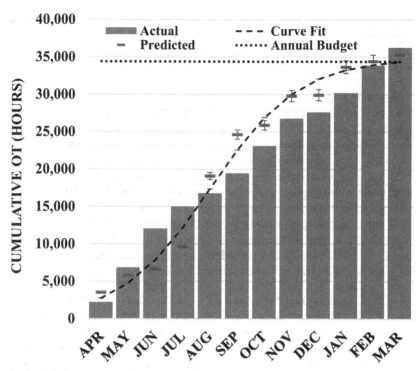

Fig. 2. Cumulative OT prediction with logistics function fit, and actual accrual over FY19/20 [4].

The COVID-19 pandemic had a significant effect on supply and demand for resources in both directions. Initially, mandates to work from home, enforced physical distancing, lack of personal protective equipment, high physical protective measures, and reluctance to increase exposure reduced levels of OT accrual. However, maintenance demands continued to accumulate, requiring personnel to complete OT to sustain operational capability. Different provinces in Canada applied policies at different points in time, leading to complications at coordinating at the federal level. Thus, it was necessary to try to incorporate change points in the system, to learn and adapt from the system dynamics as they evolved.

An attempt was made to incorporate in-year accrual data by training the system on a monthly basis to try to provide faster feedback response to the system. However, as this type of approach is dependent upon large quantities of data, new data points that are introduced are generally overwhelmed by the bulk response. As a result, it was necessary to select a new approach that could handle transient system dynamics.

2.2 Revised Approach

In this approach, since the historical OT hours are recorded by date, it was decided to examine how time series forecasting methods would perform on these data. The objective would be to estimate how the sequence of observations will continue without attempting to discover the specific factors that affect the behaviour as we did in [4]. Thereby we explore the temporal information in the same dataset from [4] in order to build a more robust predictive model. There are time series models that can extrapolate on trend and seasonal patterns, both potentially providing useful information for the comptrollers at FMF CB.

Data Pre-processing. An additional objective of the revised approach was to simplify the data pre-processing steps required. In the initial approach, two separate datasets were extracted from DRMIS. Furthermore, the month and year that WOs occurred was available in Dataset A, but the calendar date and the regular hours expended for WOs was not available in either Dataset A or B. In the revised approach, a single dataset was extracted from the official enterprise records management system which provided the expended hours (either regular or OT hours) for all WOs from FY13/14-FY20/21 by date. For each record where expended hours occurred, detailed information of maintenance specifics, financial coding and classification of personnel assigned to the work were included in order to have all data fields as were extracted in the previous approaches. The dataset included numerical, categorical and time-series data. A snapshot for all expended hours in FY21/22 was also extracted for predictive purposes.

Data cleaning were completed using the Python Pandas package [27] and filtered out any records where military personnel were assigned to the WO since the FMF overtime budget is used for civilians only. Similarly, any expended hours incurred during sea duty or sea trials were removed from the analyses, as this would not apply to second- or third-line maintenance. To build the time series models, only two fields were used; calendar date and expended OT hours (where this value was greater than zero).

Model Generation. There are a number of time series models available for forecasting data; however, given that daily data were available for several years, the Prophet model appeared to be most capable [23]. The algorithm automatically forecasts data using an additive model, where non-linear trends are fit with yearly, weekly and daily seasonality, as well as variable holiday effects. The implementation utilized is available as open-source software for both Python and R [28], with Python being used for this work.

The Prophet time-series forecast is developed using an additive regressive model, as given in Eq. 1 where it is the date and $y(t)$ is the actual OT hours accrued. The components of the model include the long-term trend, $g(t)$, the holiday effect, $h(t)$, yearly seasonality, $s_y(t)$, weekly seasonality, $s_w(t)$, and error, $e(t)$. Each of these components contains linear or non-linear terms with parameters that can be tuned to the problem at hand. In the Prophet model specification, these parameters can be adjusted manually in order to improve the model fit (referred to as Analyst-in-the-Loop modeling in [23]).

$$y(t) = g(t) + h(t) + s_y(t) + s_w(t) + e(t) \tag{1}$$

The long-term trend may be modelled as a logistic growth model or a piecewise linear model; a logistic growth model was selected in this case from a grid search used to tune the hyperparameters. The behaviour of the long-term trend can be adjusted using two hyperparameters. The first is *changepoint_ range*: this was set at 80% so that only the first 80% of the time-series data were used to infer change points in the trend to avoid overfitting. The second is *changepoint_prior_scale*, where an increase of this prior scale value adds more flexibility to the trend changes; a larger prior scale can result in the model overfitting due to too much flexibility, and a smaller prior scale can make the model not being flexible enough, causing an underfit. It was set to the recommended default value of 0.05.

The holiday effect can be tailored to country and date-specific input. Here, Canadian holidays were manually added and formatted into the Prophet data frame format using an open-source Github repository [29].

Seasonality components utilize Fourier series, with multiplicative or additive parameters for individual periods. The order of the Fourier series found to give the best fit was 3, the seasonality mode was additive, and the *seasonality prior scale* was set to 15.

Once all the hyperparameters were specified, the Python implementation of the Prophet model followed the standard *sklearn* application programming interface (API) [30]. Prophet has its own implementation of the *fit()* and *predict()* methods from the *sklearn* API; the *fit()* method was used to train the model using specified hyperparameters for the given training data and the *predict()* method was used to forecast future values for a specified period (in this case from 1 September to 31 March of the following calendar year).

3 Results

3.1 Model Fitting

The fitted model was trained on two columns of data from 01 April 2012 to 31 August 2021 with individual dates as the date series input and the daily OT hours accrued as the response; the components of the trained time series model are plotted in Fig. 3 using Python's *Matplotlib* [31] with *Seaborn* [32] to visualize the results.

The long-term trend component explains the general trend of OT hours irrespective of seasonality components. There is a slight downward trend in OT hours during last three years, which can be explained due to COVID-19 pandemic.

Holidays were found to be generally negatively correlated with OT hours, except for the Labor Day holiday, which is positively correlated.

Two different weekly seasonality are captured instead of the default weekly seasonality: 1) *weekly_on_seasonality* captures the OT patterns during regular weeks (5 January – 23 December), and 2) *weekly_off_seasonality* accounts for the effect of long holidays during Christmas and New Year's season (24 December – 4 January) where the day of week may not play as significant a role. The weekly seasonality data confirms an intuitive understanding of when OT hours are accrued – few OT hours are typically reported on weekday, while more are reported on the weekends. Saturdays see the highest number of hours reported, which may be explained as further ability and/or willingness to work Saturdays compared to Sundays.

Yearly seasonality captures the effect of the time of the year for OT hours. This seasonal data also confirms an intuitive understanding of the drop of OT hours accrued during Christmas leave and March (end of the fiscal year, likely due to restricted budgets). There are some notable peaks in the yearly seasonality component:

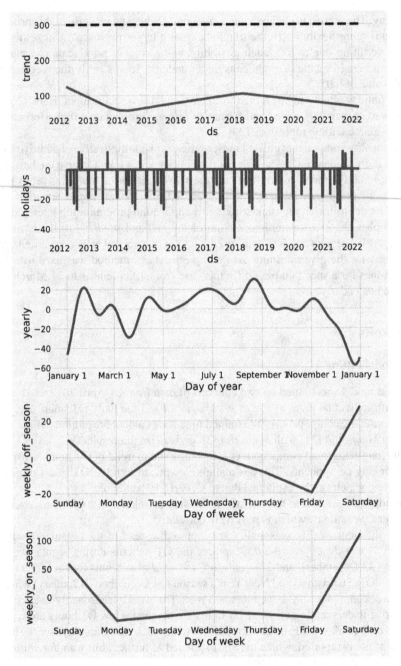

Fig. 3. Time series components of OT hours accrued for FMF CB from 01 April 2012 to 31 August 2021.

- January – after Christmas Leave, which corresponds to the start of a number of major military personnel deployments every 6 months;
- April – the beginning of new fiscal year and release of new budgets;
- July – corresponding to the start of major military personnel deployments every 6 months or summer leave season when fewer personnel are available and more OT is required to sustain operations with fewer resources; and,
- September – the end of summer leave and the start of the new military posting season.

3.2 Model Testing

The time series model was tested on partial in-year data from FMF CB from both FY19/20 and FY20/21 (Figs. 4 and 5) and compared to the actual data for the remainder of the fiscal year. The forecast was predicted from 31 August until the end of fiscal year (31 March). For each tested fiscal year, the model was only trained on the data until 31 August of that FY to avoid data leakage. Figure 4 shows the daily OT hours accrued, while Fig. 5 is the cumulative OT hours accrued.

The accuracy of the model is calculated as per Eq. 2. The cumulative forecast OT in FY19/20 was 36,683 h, and the actual OT accrued was 39,382 h; thus, the accuracy was 93.7%. The cumulative forecast OT in FY20/21 was 20,519 h, and the actual OT accrued was 22,483 h; thus, the accuracy was 90.4%.

$$Accuracy = \left(1 - \frac{|ActualOTHours - ForecastOTHours|}{ActualOTHours}\right) \times 100\% \qquad (2)$$

3.3 Model Forecasting

A forecast is given in Fig. 6 by applying the new approach to FY21/22 based on the WO available up until 31 August 2021. The total forecast OT accrued is 29,186 h.

If model is at least 90% accurate (based on the worst-case performance of the last two years of forecasts), Eq. 2 for model accuracy can be used to calculate the error interval on the total forecast estimate, as per Eq. 3. Thus, the actual OT accrued for FY21/22 is estimated to fall between 26,532 to 32,428 total hours.

$$90\% = \left(1 - \frac{|ActualOTHours - 29,186|}{ActualOTHours}\right) \times 100\% \qquad (3)$$

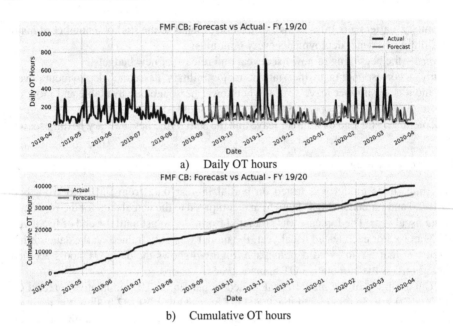

Fig. 4. Forecast versus actual OT hours accrued for FMF CB in FY19/20 (01 April 2019 – 31 March 2020).

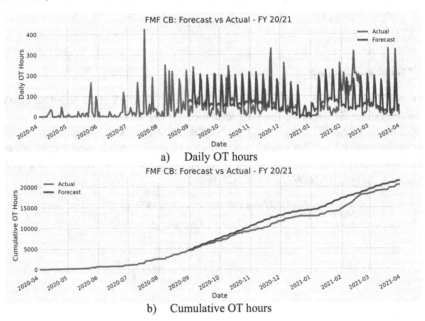

Fig. 5. Forecast versus actual OT hours accrued for FMF CB in FY20/21 (01 April 2020 – 31 March 2021).

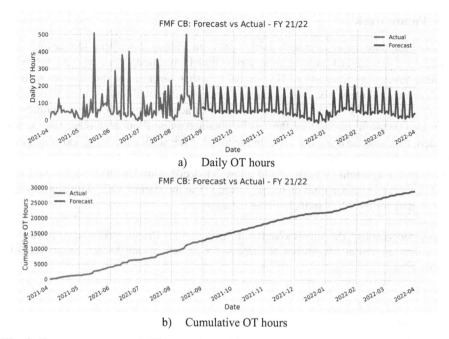

a)   Daily OT hours

b)   Cumulative OT hours

Fig. 6. Forecast versus actual OT hours accrued for FMF CB in FY21/22 (01 April 2021 – 31 March 2022).

4   Conclusions

4.1   Prior Approach

The autoML approach to forecasting OT accrual for a naval FMF was both easy to apply and accurate within the constraints of the assumptions applied. Insights to major variables of importance for OT were explored and a simple logistics function was fit for cumulative OT hours per fiscal year. However, due to the destabilizing factors introduced in March 2020 as a result of the COVID-19 pandemic, the system was no longer in steady-state operation for a variety of reasons. Attempts to refine the predictions using month-by-month forecasting and updates were not successful. As a result, an alternative method was sought to forecast OT using time-series analysis during transient system states.

4.2   Revised Approach

The time-series approach generally works well to forecast cumulative overtime hours accrued, including during large fluctuations. The advantage of the time series approach is its ability to capture different seasonality components such as holidays, day of the week effect and the time of the year, which provide more insights into the system, leading to better modeling accuracy in the predictions. Both the FY19/20 and FY20/21 models showed over 90% accuracy in predicting the total OT hours which was an improvement compared to the previous approach during transient system states.

4.3 Future Work

The revised time series approach used Prophet in Python, a scalable forecasting tool based on a generalized additive time series model. Additional areas of research may include developing forecasts of OT accrual by other variables of importance, such as work centre and employee classification. The approach could also be used to develop forecasts for FMF CS, and compare policies and accrual trends between facilities. It may also be useful to compare the results obtained using Prophet to other time series models, to determine if accuracy can be improved. After sufficient testing and validation, operationalizing these algorithms as a budgeting tool accessible from within enterprise resource management systems would allow for decision makers to have updated and readily available information on the projected OT at the start of a FY, with adjustments as new data for the FY is in included in the analysis.

Acknowledgements. The authors would like to thank the FMF Comptrollers for their availability and insight, as well as Mr. Andrew MacDonald for assistance with the intermediate autoML approach.

References

1. Royal Canadian Navy Homepage: http://www.navy-marine.forces.gc.ca/en/index.page. Accessed 10 Jan 2021
2. Holmes, M.: Predicting overtime hours for fleet maintenance facility cape breton (DRDC-RDDC-2020-R071). Defence R&D Canada - Centre for Operational Research and Analysis, Ottawa (2020)
3. h2o.ai homepage. https://www.h2o.ai/. Accessed 10 Dec 2021
4. Eisler, C., Holmes, M.: Applying automated machine learning to improve budget estimates for a naval fleet maintenance facility. In: 10th International Conference on Pattern Recognition Applications and Methods, pp. 586–593. SciTePress, Portugal (2021)
5. Maybury, D.: Predictive analytics for the royal canadian navy fleet maintenance facilities: an application of data science to maintenance task completion times (DRDC-RDDC-2018-R150). Defence R&D Canada - Centre for Operational Research and Analysis, Canada (2018)
6. Balaji, A., Allen, A.: Benchmarking automatic machine learning frameworks. https://arxiv.org/abs/1808.06492. Accessed 10 Dec 2021
7. Gijsbers, P., LeDell, E., Thomas, J., Poirier, S., Bischl, B., Vanschoren, J.: An open source AutoML benchmark. In: Proceedings of 6th ICML Workshop on Automated Machine Learning, pp. 1–8. International Conference on Machine Learning, United States. https://arxiv.org/pdf/1907.00909.pdf. Accessed 10 Dec 2021
8. He, X., Zhao, K., Chu, X.: AutoML: A Survey of the State-of-the-Art. Knowledge-Based Systems. Knowledge-Based Systems (212), (2021). https://doi.org/10.1016/j.knosys.2020.106622. Accessed 13 Oct 2021
9. Truong, A., Walters, A., Goodsitt, J., Hines, K., Bruss, C.B., Farivar, R.: Towards Automated Machine Learning: Evaluation and Comparison of AutoML Approaches and Tools. https://arxiv.org/pdf/1908.05557.pdf. Accessed 13 Oct 2021
10. Baykasoğlu, A., Öztaş, A., Özbay, E.: Prediction and multi-objective optimization of high-strength concrete parameters via soft computing approaches. Expert Syst. Appl. **36**(3, Part 2), 6145–6155 (2009)

11. Desai, R.J., Wang, S.V., Vaduganathan, M., Evers, T., Schneeweiss, S.: Comparison of machine learning methods with traditional models for use of administrative claims with electronic medical records to predict heart failure outcomes. JAMA Netw Open 3(1), e1918962 (2020)
12. Ganapathi, A., Kuno, H., Dayal, U., Wiener, J.L., Fox, A., Jordan, M., et al.: Predicting multiple metrics for queries: better decisions enabled by machine learning. In: Proceedings from 25th International Conference on Data Engineering, pp. 592–603. IEEE, China (2009)
13. Guanoluisa, D.A.Q.: Design and Implementation of a Micro-World Simulation Platform for Condition Based Maintenance using Machine Learning Algorithms. Master's thesis. University of Toronto, Canada (2020)
14. Naik, J., Dash, P.K., Dhar, S.: A multi-objective wind speed and wind power prediction interval forecasting using variational modes decomposition based multi-kernel robust ridge regression. Renewable Energy 136, 701–731 (2019)
15. Ozturk, S., Fthenakis, V.: Predicting frequency, time-to-repair and costs of wind Turbine failures. Energies 13(5), 1149 (2020)
16. De Gooijer, J., Hyndman, R.: 25 years of time series forecasting. Int. J. Forecast. 22, 443–473 (2006)
17. Ueno, R., Diener, R., Calitoiu, D.: Exploring regular force attrition with exponential smoothing: uncovering seasonality in voluntary releases. (DRDC-RDDC-2018-R214). Defence R&D Canada - Centre for Operational Research and Analysis, Canada (2019)
18. Ueno, R., Calitoui, D.: Time series methods for forecasting regular force attrition. (DRDC-RDDC-2020-R045). Defence R&D Canada - Centre for Operational Research and Analysis, Canada (2019)
19. Pilat, M., Gauthier, Y.: Fiscal Year Forecasts of Canadian Armed Forces Enrollment. (DRDC-RDDC-2019-L100). Defence R&D Canada - Centre for Operational Research and Analysis, Canada (2019)
20. Feiring, D.: Forecasting Marine Corps Enlisted Manpower Inventory Levels with Univariate Time Series Models. Master's Thesis. Naval Postgraduate School, Monterey (2006)
21. Sharma, S., Karol, S.: Modeling and forecasting of india's defense expenditures using box-Jenkins ARIMA model. Int. J. Res. 9(2), 334–344 (2021)
22. Hyndman, R.J., Athanasopoulus, G.: Forecasting: principles and practice, 3rd edn. OTexts, Melbourne, Australia (2021)
23. Taylor, S.J., Letham, B.: Forecasting at scale. Am. Stat. 72(1), 37–45 (2018)
24. Menculini, L., et al.: Comparing prophet and deep learning to ARIMA in forecasting wholesale food prices. Forecasting 3, 644–662 (2021)
25. Samal, K., Babu, K., Das, S., Acharaya, A.: Time series based air pollution forecasting using SARIMA and prophet model. In: Proceedings of the 2019 International Conference on Information Technology and Computer Communications, pp. 80–85. Association for Computing Machinery, New York (2019)
26. Jain, G., Prasad, R.R.: Machine learning, prophet and XGBoost algorithm: analysis of Traffic Forecasting in Telecom Networks with time series data. In: Proceedings of 8th International Conference on Reliability, Infocom Technologies and Optimization (Trends and Future Directions), pp. 893–897. IEEE, (2020)
27. Python Pandas Package homepage, https://pandas.pydata.org/. Accessed 13 Oct 2021
28. Prophet Forecasting at scale. https://facebook.github.io/prophet. Accessed 10 June 2021
29. University of Waterloo Holiday Dataset. https://github.com/uWaterloo/Datasets/blob/master/Holidays/holidays.csv. Accessed 15 Oct 2021
30. sklearn Github Repository. https://github.com/scikit-learn/scikit-learn. Accessed 15 Oct 2021
31. Matplotlib: Visualization with Python. https://matplotlib.org/. Accessed 15 Oct 2021
32. seaborn: statistical data visualization. https://seaborn.pydata.org/index.html. Accessed 15 Oct 2021

Exploiting Temporal Coherence to Improve Person Re-identification

Oliverio J. Santana[✉][iD], Javier Lorenzo-Navarro[iD], David Freire-Obregón[iD],
Daniel Hernández-Sosa[iD], José Isern-González[iD], and Modesto Castrillón-Santana[iD]

SIANI, Universidad de Las Palmas de Gran Canaria, Las Palmas de Gran Canaria, Spain
oliverio.santana@ulpgc.es

Abstract. The uncontrolled characteristics of long-term scenarios, like ultra-running competitions, are challenging for person re-identification approaches based on computer vision methods. State-of-the-art techniques have reported hardly moderate success for whole-body runner re-identification due to the existence of distinct illumination conditions, as well as changes of clothing and/or accessories like backpacks, caps, and sunglasses. This paper explores integrating these biometric cues with the particular spatio-temporal context information present in the competition live track system. Our results confirm the significance of this strategy to limit the gallery size and boost re-identification performance.

Keywords: Temporal coherence · Ultra-distance race · Sporting event · Person re-identification · Computer vision

1 Introduction

Nowadays, running events of all kinds are becoming increasingly popular, opening up various business opportunities. The organizers of massive running competitions, attended by hundreds or even thousands of participants, face significant logistical challenges. Among these challenges, managing the race ranking is the most relevant in the long term.

To control each runner's position in the race ranking, personal chips are generally used to identify the participants as they pass through each track checkpoint. Based on the data provided by these chips, rankings are generated and prizes are awarded. Since only particular locations are monitored, it would be possible for a runner to cheat by taking shortcuts and not following the entire course of the race. Still, the timing system permits the analysis of all the collected data to determine whether a participant is suspicious of course cutting.

This work is partially funded by the ULPGC under project ULPGC2018-08, by the Spanish Ministry of Economy and Competitiveness (MINECO) under project RTI2018-093337-B-I00, by the Spanish Ministry of Science and Innovation under projects PID2019-107228RB-I00 and PID2021-122402OB-C22, and by the ACIISI-Gobierno de Canarias and European FEDER funds under projects ProID2020010024, ProID2021010012, ULPGC Facilities Net, and Grant EIS 2021 04.

M. De Marsico et al. (Eds.): ICPRAM 2021/2022, LNCS 13822, pp. 134–151, 2023.
https://doi.org/10.1007/978-3-031-24538-1_7

An evident flaw of this approach is that it is not the person but the chip being monitored. In addition, there is no control over whether the runner carrying the chip tag is the same person who registered for the race. Indeed, there are no mechanisms to verify whether the runner who passes through each checkpoint carrying a particular chip is always the same person. Such a situation would create substantial issues for the organization, especially regarding insurance policies, while the bad user experience generated by wrong ranking results would damage the reputation of the organizers and the race event itself.

Deep learning and computer vision strategies can be used to address these problems. Some initial proposals have focused on detecting the racing bib number [1], although this would not solve the aforementioned issues. Thus it is necessary to look for biometric features to identify the person wearing the bib [5, 18, 23].

In this context, our work focuses on the re-identification (ReID) of runners participating in an ultra-marathon race. We capture the images of a runner passing through a particular checkpoint (probe) and verify if they belong to the same person who passed through one of the previous checkpoints (gallery). This is a real-world wild scenario in which any computer vision method would struggle due to the wide variety of runner poses, as well as the challenges raised by scene lighting, camera location, image resolution, sharpness, focus, motion blur, etc.

In our previous work [14], we explored the integration of different elements to boost ReID performance in this complex scenario: face recognition, body recognition, and temporal coherence. With the latter, the size of the gallery for a probe runner in a particular checkpoint is significantly reduced by filtering out those runners who have already left that checkpoint and those runners who have not had enough time to reach it. This is achieved by making use of the temporal information available to the race organization at the moment a runner passes the probe checkpoint.

Our original temporal coherence strategy does indeed boost ReID performance, but it relies on an adaptive threshold computed for each runner according to their progression through the previous stages of the race. Although this heuristic proves helpful in the later stages of the race, its improvement is limited in earlier stages, as there is less information available to estimate the appropriate threshold value.

In this new work, we analyze different strategies for improving ReID using time coherence. In addition, since face recognition has proven to yield poor ReID performance in this scenario, we focus on using only techniques based on the body appearance. Our results reveal significantly better performance than the one reported in our previous paper, emphasizing the importance of considering all the available context information for ReID tasks.

2   Related Work

Computer vision in sporting events is a challenging research field, with an active community as suggested by recent surveys [15, 21] and regular conference workshops such as CVPR's CVsports and ACM's MMSports. In those venues, the community has mainly focused on the analysis of athlete movements and the study of team sports to collect statistical data and evidence different ways for improvement [10, 21]. Since this work is focused on ReID for race participants, we briefly summarise the most relevant proposals on this topic.

Race bib number recognition is the most commonly adopted approach for runner ReID. Most works in this area deal with the marathon context, characterized by daylight conditions and large bib number fonts. However, even when these competitions are being organized almost everywhere, the amount of publicly available data is limited. The pioneering work by Ben-Ami et al. [1] encloses a public dataset for race bib number detection and recognition. Firstly, face detection is applied to estimate the region of interest, i.e., the most likely bib position, to later identify the bib number using optical character recognition. Shivakumara et al. [20] and Boonsim [2] also adopt the initial face detection step to later apply text detection and recognition. On the other hand, De Jesus and Borges [8] skip the first face detection step, focusing on text detection, but at the risk of getting confused by runner poses.

More recently, deep learning has been adopted for bib number recognition. Kamlesh et al. [9] use TextBoxes [11] to detect the bib location and then a convolutional recurrent neural network to identify the runner according to the bib number, essentially simplifying the person ReID problem into a text recognition problem. Wong et al. [22] propose a similar two-stage approach, where the popular *You Only Look Once* algorithm [19] is applied to detect the runner and the bib, while a convolutional recurrent neural network is later used to recognize the number. A more complex multi-stage approach is proposed by Nag et al. [16], using the single shot multibox detector [12] to detect individuals and then extracting those body parts more likely to contain the bib; the bib is later detected by means of a convolutional neural network and the number is finally identified using a convolutional recurrent neural network.

The main drawback of runner ReID using the bib number is that the digits cannot be guaranteed to be clearly visible, as illustrated in Fig. 1. Ultra-distance runners tend to prioritize comfort and may wear the race bib in positions where capturing a clear image becomes difficult, if not impossible. When the bib number is not a reliable ReID method, it would be necessary to use the information provided by more visual cues, such as the face, body, or clothing. However, this approach is much less frequent in the literature.

Wrońska et al. [23] combine facial appearance with race bib number recognition to improve the overall ReID performance. This double identification improves performance in marathon-like scenarios captured in daylight conditions, but it has problems detecting features on darker images. Ultra-distance races provide more challenging in the wild scenarios. Indeed, face and bib number occlusions are frequent due to the runner poses. Bearing this in mind, Peñate et al. [18] define a new benchmark for ultra-marathon ReID approaches based on the body/clothing appearance. The evaluation presented in that work for top-ranked ReID approaches shows the scenario difficulties, mainly because the race covers a significant time lapse and participants may change their clothes along the track due to the weather conditions or simply for personal hygiene.

Fig. 1. Runners with the race bib not placed in front or folded.

The inherent low resolution of the faces captured during an ultra-distance race makes face ReID provide significantly worse results than face recognition techniques in other scenarios. As pointed out by Cheng et al. [4], differences exist in the classical surveillance scenario compared to the familiar face recognition context, further evidenced by the reduced success obtained by present face-based techniques in challenging benchmarks. This is also shown by Dietlmeier et al. [7], suggesting the reduced negative effect of blurring faces in ReID benchmarks regarding overall system accuracy.

Interestingly, all of the aforementioned proposals are based on photographs or static frames extracted from videos. Runner ReID using the dynamic information in video footage is getting more attention lately. We may mention the recent CampusRun dataset [17] of images captured for each participant every kilometer by hand-held cameras during a half marathon event, offering a large number of samples per identity. Given this video stream availability, the gait trait is adopted by Choi et al. [5] focusing on the arm swing features extracted from the silhouette. The author's strategy is to remove the problems present in race bib number or face occlusions, as well as the similarity in clothing appearance, e.g., runners of the same team. Another interesting scope focuses on player ReID in broadcast videos of team sports. For instance, Comandur [6] has developed a hierarchical data sampling procedure and a centroid loss function, increasing ReID performance when combined.

3 Dataset

We evaluate ReID performance using a dataset of images collected during Transgrancanaria 2020 [18], a mountain ultra-marathon race in which runners face a wide variety of extremely demanding conditions. It is a 128 km race that takes runners across the island of Gran Canaria from north to south. The 2020 race edition started at 11 pm on March 6th (one week before the Covid-19 lockdown in Spain), and the finish line closed 30 h later, although the winners only needed 13 h to complete the entire distance. The information provided by this dataset is complemented with the official rankings published by the race organizer, Arista Eventos SLU, to evaluate the impact of time coherence strategies on ReID performance.

Runners were captured at five recording points (RPs) distributed along the race track. Images were recorded using a set of Sony Alpha ILCE 6400 (16–50 mm lens) configured at 50 fps and 1920 × 1080 pixel resolution. Figure 2 shows the recording setup near some of the RPs.

Fig. 2. Recording setup at different RPs.

The capture conditions vary significantly among the different RPs. The first two RPs were located in the first 30 km, while the last three were located in the last 40 km, see Table 1. Those locations affect the starting recording time. Indeed the first two RPs were recorded at nighttime and the remaining three were recorded in daylight. Figure 3 shows a collage of sample images taken at the race start and each of the five RPs. It becomes clear that there are significant differences, not just among them. Still, it is evident that frames extracted from video footage differ from those taken by a professional. Figure 4 shows professional photographs with well-focused and posed runners, sometimes even looking at the camera. Actual footage is far from ideal snapshots, hence the great difficulty of ReID tasks in wild scenarios.

Table 1 also summarizes the dataset statistics. Due to the arduous nature of the race, only 435 of the 677 runners who started the race completed the course within the time limit. The closer to the finish line, the lower the number of annotated participants, even if the number of captured frames is larger because the elapsed time between each runner increases as the race progresses. To keep this ReID analysis within a closed set of identities, our results have been obtained considering only the 109 runners recorded and annotated in all the five RPs.

Race Start: Playa de Las Canteras

RP1: Arucas

RP2: Teror

RP3: Presa de Hornos

RP4: Ayagaures

RP5: Parque Sur

Fig. 3. Leaders of the Transgrancanaria Classic 2020 recorded at the different RPs. Figure from [14], original images from [18].

Fig. 4. The ideal case: images captured by a professional photographer during Transgrancanaria 2014. Photographs courtesy of Carlos Díaz-Recio and Arista Eventos SLU.

Table 1. Dataset statistics: location in the race track (kilometer point), recording start time (Saturday, March 7th, 2020), total number of frames in the recorded footage, and number of annotated runners. Data from [18].

	Km	Start time	Footage (frames)	Runners
RP1	16.5	00:06	140,616	419
RP2	27.9	01:08	432,624	586
RP3	84.2	07:50	667,872	203
RP4	110.5	10:20	1,001,208	139
RP5	124.5	11:20	1,462,056	114

4   Proposal Description

We adopt a ReID scenario instead of a verification scenario because multiple runners may reach a RP together, not being straightforward for the system to relate a timing system tag with a captured individual. The classic ReID experimental scenario aims to determine which runner in a gallery matches the probe identity. For each runner image, the dataset provides a bounding box enclosing its whole body. In order to perform the ReID process, runners must be represented as numerical vectors providing body descriptors. Two recent ReID frameworks were evaluated for the chosen scenario in [18] to generate these embeddings, AlignedReID++ [13] and ABD-Net [3], since they have provided state-of-the-art results in the popular Market-1501 benchmark [24]. In this work, we have chosen to use the AlignedReID++ framework to compute the embeddings because it provides slightly better performance according to [13] .

Our work relies on the fact that, besides body embeddings, the ReID process can be supported by contextual information. An ultra-running competition has a clear timeline, as participants reach each checkpoint in order, not repeating any of them. The proposed vision-based system would work in parallel with traditional timing systems, i.e., the elapsed times of every runner are available at each location.

Lets t be the total number of RPs in the race track, we consider an individual detected in RP_p as probe, where p belongs to the interval $[2, t]$, and all the samples from RP_g as the gallery, where g belongs to the interval $[1, p - 1]$. Note that any RP_g

is always preceding RP_p, and thus a runner captured in RP_1 can never be the probe. Assuming there are n_k participants recorded in RP_k, where k is p or g, we call $r_{i,k}$ to the runner crossing in position *ith* through RP_k and $t_{i,k}$ his/her elapsed time. Hence, given the probe sample $r_{i,p}$ in RP_p, its gallery in RP_g is defined as:

$$gallery = \{r_{j,g}\} \text{ for } j = 1, \ldots, n_g \tag{1}$$

However, not all runners in the gallery are coherent ReID candidates. On the one hand, we can exclude from the gallery those runners that have already crossed RP_p, that is, to identify the probe runner $r_{i,p}$ we can discard from the gallery the runners $r_{1,p}$ to $r_{i-1,p}$. In other words, we filter out those runners who already left RP_p.

On the other hand, it is possible that not every runner captured in RP_g had enough physical time to reach RP_p by the time the probe runner crossed RP_p. If we can predict how long it will take for a runner who has crossed RP_g to reach RP_p, we can determine whether the runner may have reached RP_p to exclude him/her from the gallery. In the following subsections, we discuss different strategies, extending those described in [14], to filter out those runners who could not have arrived at RP_p.

4.1   Runners in the Gallery

As illustrated in Fig. 5, when the first runners arrive at RP_p, some runners have still not crossed through RP_g. Taking this into account, the most simple temporal coherence heuristic for filtering the gallery of a probe runner $r_{i,p}$ would be to remove those gallery samples that have not yet reached RP_g.

$$gallery = \{r_{j,g}\} \text{ such that } t_{j,g} < t_{i,p} \tag{2}$$

4.2   Fastest Elapsed Time

Given the spatial distances between RPs, any runner would need some time to reach RP_p after crossing RP_g physically. Thus, we can apply more restrictive filtering establishing that runners should have passed RP_g at least x minutes ago. A possible value for x would be the organization's expected time between RPs, estimated in advance based on the distance and accumulated positive slope between RPs. However, personal issues and weather conditions may affect runner performance. As a conservative approach, we calculate the fastest possible time between RP_p and RP_g as the difference between the elapsed time of the first runner to cross both RPs. Thus, we filter the gallery assuming that any runner should need at least that time to cover the distance between both points.

$$\Delta t = t_{1,p} - t_{1,g}$$
$$gallery = \{r_{j,g}\} \text{ such that } t_{j,g} \leq t_{i,p} - \Delta t \tag{3}$$

Nevertheless, we should take into account that runners could exchange positions along the race. We illustrate this circumstance with an example. Consider that the first three runners depicted in Fig. 5 (red, blue, and green t-shirts) passed RP_g at times $00:05:00$, $00:05:20$, and $00:06:30$, respectively. The same three runners leaded the race in RP_p but exchanged positions (green, blue, and red), passing at times $01:15:32$, $01:15:50$, and $01:16:30$, respectively.

If we are considering the gallery set for the first probe runner $r_{1,p}$ then candidate runners in the gallery should verify $t_{j,g} \leq t_{1,p} - \Delta t$ which sets the temporal threshold to $01:15:32 - 00:05:00 = 01:10:32$, excluding runners blue and green from the gallery, i.e. the desired runner will be excluded from his own gallery. To overcome that circumstance, we weight the temporal threshold introducing a value thr, where $thr \in (0,1]$, to avoid removing an excessive number of identities from the gallery.

$$gallery = \{r_{j,g}\} \text{ such that } t_{j,g} \leq t_{i,p} - (\Delta t \times thr) \tag{4}$$

Setting the thr value to 1.0 would leave the previous equation unchanged, but some runners may be excluded from their gallery. Assigning a lower value will define a margin to skip those situations. We have found that a thr value of 0.9 is enough to prevent the probe runner from being excluded from the gallery in all cases, since the RPs in this dataset are located at least ten kilometers away from each other, with the fastest runners requiring almost one hour to cover the distance.

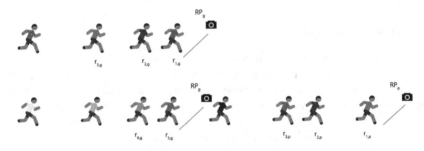

Fig. 5. Temporal coherence illustration. Top) Let's assume that red, blue, and green runners cross RP_g in that order. Bottom) Red and green runners exchange positions in RP_p when they arrive roughly one hour later. Using temporal coherence, for a probe runner arriving at RP_p, we remove from the gallery those individuals who have not had enough time to reach RP_p from RP_g. Figure adapted from [14] (Color figure online).

4.3 Adaptive Time Estimation

Regarding the real race, the fastest split time is not reachable for most runners. Due to the conservative nature of a fixed threshold, we have explored using a variable threshold defined by the runner performance. The elapsed time of probe runner $r_{i,p}$ in the previous points RP_1 to RP_{p-1} serve to estimate the elapsed time from RP_{p-1} to RP_p.

$$\widetilde{\Delta t_{i,p}} = w_0 + w_1 \Delta t_{i,2} + \cdots + w_{p-2} \Delta t_{i,p-1} \tag{5}$$

We evaluated such estimation using linear regression and random forest models. The latter reported better results and therefore was adopted to compute the adaptive threshold to filter the gallery for a particular runner $r_{i,p}$.

$$gallery = \{r_{j,g}\} \text{ such that } t_{j,g} \leq t_{i,p} - (\widetilde{\Delta t_{i,k}} \times thr) \tag{6}$$

4.4 Average Time

Accurately estimating each runner's performance requires using as much information as possible. This would be a problem for the RPs located at the beginning of the race because less information is available. To address this problem, we consider a second adaptive mechanism: instead of estimating each runner's performance, we assess all runners' overall performance. As runners arrive at RP_p, we get their elapsed time from the live track system. Then we can calculate and update the average time that all the runners have needed to reach RP_p from RP_g.

$$gallery = \{r_{j,g}\} \text{ such that } t_{j,g} \le t_{i,p} - (\frac{\sum_1^{i-1} \Delta t}{i-1} \times thr) \tag{7}$$

This average value considers the race's particular conditions, as the weather will affect all runners alike. It is also updated as the race progresses, as runners will take longer to get from one point to the next. However, there will be fluctuations in this value, and there may also appear some runners who go faster than the average at any given time, which is why we have found it necessary to use a more aggressive thr value of 0.85.

4.5 Individual Time

When a runner reaches a particular RP, the race timing system detects the chip in the bib and enters the elapsed time in the database. As this system works in parallel with runner ReID, we can take advantage of this information to apply a more restrictive strategy: filtering the gallery according to the time the runner has spent traveling the distance from RP_g to RP_p, that is, removing all the runners who have not been able to arrive in the time measured for the probe runner.

$$gallery = \{r_{j,g}\} \text{ such that } t_{j,g} \le t_{i,p} - \Delta t_{i,p} \tag{8}$$

Since the runners who have already left the RP_p are also removed, the gallery will consist only of the runner him/herself and those who should have arrived before him/her and have not yet arrived, i.e., those runners that the probe runner has passed on his way from RP_g. In addition, as we know exactly how long it took the runner to cover the distance, it is not necessary to weigh this value with a thr factor because the probe runners cannot be excluded from their galleries.

5   ReID Experimental Evaluation

To compare the different approaches to match the runners embeddings, we use the mean average precision (mAP) score, a metric well established in the recent ReID literature [3, 13], especially when each identity may be present more than once in the probe and gallery sets. Given a probe set with a total of n_p runners, the average precision AP_i for each probe runner i is computed as the area under the precision-recall curve for that runner. In particular, we compute the AP_i value using the euclidean distance, since we have checked other distance functions and found no significant variation. Once AP_i has been calculated for every runner, the mAP value is computed as the arithmetic mean of the AP_i for each probe runner.

$$mAP = \frac{\sum_{i=1}^{n_p} AP_i}{n_p} \tag{9}$$

Table 2 shows the mAP results using as probe all the RPs except the first one. For each RP_p, all the previous RPs are used separately as gallery. As expected, the RPs with nighttime images show evident poor results, although the worst effects are demonstrated by RP3, which is recorded at daybreak in a region with broad shadows. The best results appear in the daytime RPs, although there is considerable room for improvement, illustrating the great difficulty of this scenario.

Going deeper, the first column of Table 2 shows the mAP results without applying any temporal coherence method, and the second column shows the mAP results when using the most straightforward temporal coherence technique: removing from the gallery the runners who already left RP_p. The significant improvements achieved with this simple rule show the potential of temporal coherence.

Table 2. ReID mAP without applying any temporal coherence method (base) and removing the runners who already left the probe RP (left) from the gallery. The *left* method was proposed in [14] but not evaluated separately.

	Gallery	Base	Left
RP2	RP1	15.0	25.3
RP3	RP1	9.2	17.7
	RP2	7.2	16.7
RP4	RP1	23.0	35.6
	RP2	13.6	27.1
	RP3	19.5	25.9
RP5	RP1	22.7	35.4
	RP2	20.6	31.3
	RP3	15.2	21.7
	RP4	48.4	57.3

Table 3 shows the mAP results using the alternatives proposed to control the runners that cannot have arrived at RP_p. All these methods are mutually exclusive, though we have included for all of them the rule that removes the runners who have already left RP_p. In this sense, we are effectively addressing the two extremes of the time interval: those runners who had no time to arrive and those who had already left.

Removing the runners that have not reached RP_g has little impact in the earlier RPs because, by the time the first of the captured runners reaches RP_p, the last of the captured runners has already passed by RP_g. The effect is only noticeable in the later RPs, especially in RP5 when RP4 is used as a gallery, since the runners are already far apart at that point in the race. Removing the runners that have not beaten the fastest possible time between RP_p and RP_g provides better results. Still, the improvement is slight, especially in the initial RPs where runners are closer.

The adaptive strategy does not involve a significant performance improvement. It even implies a performance degradation in the initial RPs. There is not enough information on each runner's behavior at these earlier points. Thus, the elapsed time estimation provides poor results. The strategy that calculates the average elapsed time alleviates this problem since the assessment does not depend on the individual behavior of each runner. However, this strategy still does not provide significant performance improvements.

Table 3. ReID mAP removing from the gallery the runners who already left the probe RP (base) and the runners who cannot have arrived: runners who did not reach the gallery RP (gallery), runners who had not beaten the fastest possible time (fastest), runners who had not beaten the estimated time for the probe runner (adaptive), runners who had not beaten the average time for the probe RP (average), and runners who had not beaten the particular time of the probe runner (individual). The *gallery*, *fastest*, and *adaptive* methods were proposed in [14], but only results for the latter were provided. The *average* and *individual* methods are new contributions.

	Gallery	Base	Gallery	Fastest	Adaptive	Average	Individual
RP2	RP1	25.3	25.3	26.6	17.9	26.6	63.6
RP3	RP1	17.7	17.7	17.7	16.0	17.7	48.6
	RP2	16.7	16.7	16.8	10.3	16.7	47.3
RP4	RP1	35.6	35.6	35.6	36.3	35.6	61.9
	RP2	27.1	27.1	27.1	28.2	27.1	59.8
	RP3	25.9	26.8	32.5	39.2	35.5	77.2
RP5	RP1	35.4	35.4	35.4	36.8	35.4	61.8
	RP2	31.3	31.3	31.3	33.0	31.3	61.5
	RP3	21.7	21.8	26.4	28.4	26.7	71.8
	RP4	57.3	65.6	76.1	78.7	83.2	97.4

Overall, the heuristic that provides best results is the one considering the particular time of each individual runner, since it is the most restrictive approach. We analyze the cumulative matching characteristics curve (CMC) to provide more insight. This metric ranks the gallery samples according to the distance to the probe. For a given probe, the CMC rank-k accuracy is 100% if the first k ranked gallery samples contain the probe identity and 0% otherwise. The final CMC curve averages the respective probe curves. For instance, a CMC rank-1 of 80% indicates that the correct identity is ranked first for 80% of the probes.

Figure 6 shows the CMC curves for ReID without time coherence applied, ranging from rank-1 to rank-10. Once again, the great difficulty of this in the wild ReID problem becomes evident. In most cases, when runner ReID is not successful, the correct identity is not even in the top ten. The only curve that shows good behavior involves the two RPs recorded during daylight, i.e., RP4 (gallery) and RP5 (probe).

RP2 probe

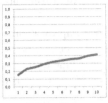

RP1 gallery

RP3 probe

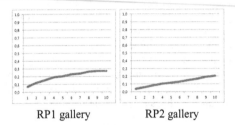

RP1 gallery   RP2 gallery

RP4 probe

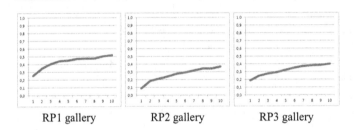

RP1 gallery   RP2 gallery   RP3 gallery

RP5 probe

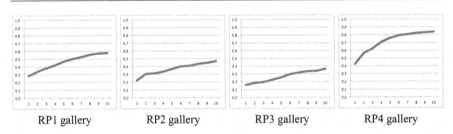

RP1 gallery   RP2 gallery   RP3 gallery   RP4 gallery

Fig. 6. CMC curves for ReID without time coherence: all the identities captured in RP_g are used as gallery. The horizontal axis represents ranks 1 to 10. The vertical axis shows the cumulative value from 0 to 1.

RP2

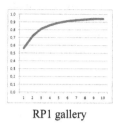

RP1 gallery

RP3

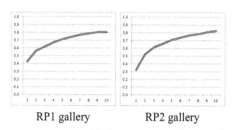

RP1 gallery   RP2 gallery

RP4

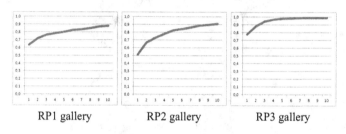

RP1 gallery   RP2 gallery   RP3 gallery

RP5

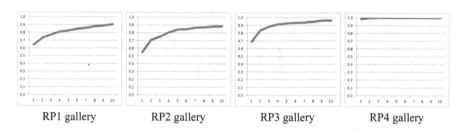

RP1 gallery   RP2 gallery   RP3 gallery   RP4 gallery

Fig. 7. CMC curves for ReID using the best time coherence strategy: excluding those runners that cannot have arrived to RP_p because they had not beaten the probe runner elapsed time and those runners who already left RP_p. The horizontal axis represents ranks 1 to 10. The vertical axis shows the cumulative value from 0 to 1.

Figure 7 shows the CMC curves for ReID using the best time coherence strategy. All the curves show good behavior; in most cases the correct identity is classified in the top ten. Indeed, in the full daylight scenario, involving RP5 as the probe and RP4 as the galleries, all the identities are classified at least as rank-3. Given the high performance of the classifier in this RP, it is interesting to examine the reasons behind the identification errors still occurring.

Figure 8 shows an example of a misidentified runner in this probe-gallery combination. As can be observed, there is some physical resemblance between the runners, and they are wearing similar clothing: sneakers, shorts, and a backpack. The most notable differences are in the cap and the shirt, although the shadows in the gallery images make it difficult to perceive the color contrast of the latter. However, the most remarkable detail is that the probe runner was not wearing the bib when he passed through the gallery RP, which has undoubtedly been a decisive factor in confusing the classifier. These results prove that, even when we can use contextual information to reduce the gallery size, there is still room for improvement.

probe runner   wrong gallery   correct gallery

Fig. 8. Misidentified runner using RP5 as probe and RP4 as gallery.

Fig. 9. The same subject captured in RP3 and RP4. Medium or long-term ReID implies that subjects will not only vary in the location and time they are captured, but also their clothes may change.

6   Conclusions

Addressing in the wild scenarios poses challenging difficulties for current computer vision methods. Using datasets with well-illuminated individuals and trained poses is helpful for testing the theoretical potential of a given method, but the reality is far from ideal. Therefore, it is is also essential to analyze the behavior of these methods with more complex datasets, closer to the actual conditions that any video surveillance system would have to face. Long-distance races offer the opportunity to create such datasets, where there are not only significant changes in lighting and landscape but also the physical appearance of the runners varies with changes in clothing, not to mention changes in the position of the race bib.

As evidenced in Fig. 9, the possibility that runners may change their clothes, given the time gap among RPs, makes ReID much more challenging when it is grounded uniquely on body appearance. Certainly, our earlier evaluations [14,18] and the literature on face recognition at this resolution [4,7] do not help much to cope with the problem. It seems that other cues must be considered in the future to make the problem completely tractable, such as monitoring gait or focusing on detailed human body parsing to analyze specific elements of the runner outfit.

In our previous paper [14], we proposed to use the contextual information provided by the live track timing system to refine the runner gallery and boost ReID performance by excluding all those identities that would be inconsistent for the probe runner at a particular point of the race track, either because the corresponding runner has already passed that point or because he/she had no material time to get there. This new work revisits the time coherence strategies proposed in [14] and presents new ones that achieve significantly better performance.

Our best time coherence strategy provides near-optimal results at the points where the probe runner and the gallery samples are captured in good lighting conditions. However, there is plenty of room for improvement on the rest of the points. Having datasets in which current techniques cannot identify runners, not even using temporal information, provides an interesting testing ground for the new computer vision methods that will be developed in the future.

Acknowledgements. We would like to thank Arista Eventos SLU and Carlos Díaz Recio for granting us the use of Transgrancanaria media. We would also like to thank the volunteers and researchers who have taken part in the data collection and annotation, as well as the previous papers of this project.

References

1. Ben-Ami, I., Basha, T., Avidan, S.: Racing bib numbers recognition. In: Proceedings of the British Machine Vision Conference, pp. 19.1–19.10 (2012). https://doi.org/10.5244/C.26.19
2. Boonsim, N.: Racing bib number localization on complex backgrounds. WSEAS Trans. Syst. Control **13**, 226–231 (2018)
3. Chen, T., et al.: ABD-Net: attentive but diverse person re-identification. In: 2019 IEEE/CVF International Conference on Computer Vision (ICCV), pp. 8350–8360 (2019). https://doi.org/10.1109/ICCV.2019.00844
4. Cheng, Z., Zhu, X., Gong, S.: Face re-identification challenge: are face recognition models good enough? Pattern Recogn. **107**, 107422 (2020). https://doi.org/10.1016/j.patcog.2020.107422
5. Choi, Y., Napolean, Y., van Gemert, J.C.: The arm-swing is discriminative in video gait recognition for athlete re-identification. In: 2021 IEEE International Conference on Image Processing (ICIP), pp. 2309–2313 (2021). https://doi.org/10.1109/ICIP42928.2021.9506348
6. Comandur, B.: Sports Re-ID: improving re-identification of players in broadcast videos of team sports (2022). https://doi.org/10.48550/ARXIV.2206.02373
7. Dietlmeier, J., Antony, J., McGuinness, K., O'Connor, N.E.: How important are faces for person re-identification? In: 2020 25th International Conference on Pattern Recognition (ICPR), pp. 6912–6919 (2021). https://doi.org/10.1109/ICPR48806.2021.9412340
8. de Jesus, W.M., Borges, D.L.: An improved stroke width transform to detect race bib numbers. In: Mexican Conference on Pattern Recognition, pp. 267–276 (2018). https://doi.org/10.1007/978-3-319-92198-3_27
9. Kamlesh, Xu, P., Yang, Y., Xu, Y.: Person re-identification with end-to-end scene text recognition. In: Chinese Conference on Computer Vision, pp. 363–374 (2017). https://doi.org/10.1007/978-981-10-7305-2_32
10. Li, G., Zhang, C.: Automatic detection technology of sports athletes based on image recognition technology. EURASIP J. Image Video Process. **2019**(1), 1–9 (2019). https://doi.org/10.1186/s13640-019-0415-x
11. Liao, M., Shi, B., Bai, X., Wang, X., Liu, W.: TextBoxes++: a fast text detector with a single deep neural network. In: Proceedings of the Thirty-First AAAI Conference on Artificial Intelligence (AAAI-2017), pp. 4161–4167 (2017)

12. Liu, W., et al.: SSD: single shot multibox detector. In: Leibe, B., Matas, J., Sebe, N., Welling, M. (eds.) ECCV 2016. LNCS, vol. 9905, pp. 21–37. Springer, Cham (2016). https://doi.org/10.1007/978-3-319-46448-0_2

13. Luo, H., Jiang, W., Zhang, X., Fan, X., Qian, J., Zhang, C.: AlignedReID++: dynamically matching local information for person re-identification. Pattern Recogn. **94**, 53–61 (2019). https://doi.org/10.1016/j.patcog.2019.05.028

14. Medina, M.A., Lorenzo-Navarro, J., Freire-Obregón, D., Santana, O.J., Hernández-Sosa, D., Castrillón-Santana, M.: Boosting re-identification in the ultra-running scenario. In: Proceedings of the 11th International Conference on Pattern Recognition Applications and Methods (ICPRAM), pp. 461–469 (2022). https://doi.org/10.5220/0010904600003122

15. Moeslund, T.B., Thomas, G., Hilton, A. (eds.): Computer Vision in Sports. ACVPR, Springer, Cham (2014). https://doi.org/10.1007/978-3-319-09396-3

16. Nag, S., Ramachandra, R., Shivakumara, P., Pal, U., Lu, T., Kankanhalli, M.: CRNN based jersey-bib number/text recognition in sports and marathon images. In: 2019 International Conference on Document Analysis and Recognition (ICDAR), pp. 1149–1156 (2019). https://doi.org/10.1109/ICDAR.2019.00186

17. Napolean, Y., Wibowo, P.T., van Gemert, J.C.: Running event visualization using videos from multiple cameras. In: Proceedings of the 2nd International Workshop on Multimedia Content Analysis in Sports, pp. 82–90 (2019). https://doi.org/10.1145/3347318.3355528

18. Penate-Sanchez, A., Freire-Obregón, D., Lorenzo-Melián, A., Lorenzo-Navarro, J., Castrillón-Santana, M.: TGC20ReId: a dataset for sport event re-identification in the wild. Pattern Recogn. Lett. **138**, 355–361 (2020). https://doi.org/10.1016/j.patrec.2020.08.003

19. Redmon, J., Divvala, S., Girshick, R., Farhadi, A.: You only look once: unified, real-time object detection. In: 2016 IEEE Conference on Computer Vision and Pattern Recognition (CVPR), pp. 779–788 (2016). https://doi.org/10.1109/CVPR.2016.91

20. Shivakumara, P., Raghavendra, R., Qin, L., Raja, K.B., Lu, T., Pal, U.: A new multi-modal approach to bib number/text detection and recognition in marathon images. Pattern Recogn. **61**, 479–491 (2017). https://doi.org/10.1016/j.patcog.2016.08.021

21. Thomas, G., Gade, R., Moeslund, T.B., Carr, P., Hilton, A.: Computer vision for sports: current applications and research topics. Comput. Vision Image Underst. **159**, 3–18 (2017). https://doi.org/10.1016/j.cviu.2017.04.011

22. Wong, Y.C., Choi, L.J., Singh, R.S.S., Zhang, H., Syafeeza, A.R.: Deep learning based racing bib number detection and recognition. Jordanian J. Comput. Inf. Technol. (JJCIT) **5**(3), 181–194 (2019). https://doi.org/10.5455/jjcit.71-1562747728

23. Wrońska, A., Sarnacki, K., Saeed, K.: Athlete number detection on the basis of their face images. In: 2017 International Conference on Biometrics and Kansei Engineering (ICBAKE), pp. 84–89 (2017). https://doi.org/10.1109/ICBAKE.2017.8090642

24. Zheng, L., Shen, L., Tian, L., Wang, S., Wang, J., Tian, Q.: Scalable person re-identification: a benchmark. In: 2015 IEEE International Conference on Computer Vision (ICCV), pp. 1116–1124 (2015). https://doi.org/10.1109/ICCV.2015.133

Perusal of Camera Trap Sequences Across Locations

Anoushka Banerjee[✉], Dileep Aroor Dinesh, and Arnav Bhavsar

MANAS Lab, SCEE, Indian Institute of Technology Mandi, Kamand, H.P, India
anoushkadeer@gmail.com, {addileep,arnav}@iitmandi.ac.in

Abstract. The current rate of decline in biodiversity exclaims ecological conservation. In response, camera traps are being increasingly deployed for the perlustration of wildlife. The analyses of camera trap data can aid in curbing species extinction. However, a substantial amount of time is lost in the manual review curtailing the usage of camera traps for prompt decision-making. The insuperable visual challenges and proneness of camera trap to record empty frames (frames that are natural backdrops with no wildlife presence) deem wildlife detection and species recognition a demanding and taxing task. Thus, we propose a pipeline for wildlife detection and species recognition to expedite the processing of camera trap sequences. The proposed pipeline consists of three stages: (i) empty frame removal, (ii) wildlife detection, and (iii) species recognition and classification. We leverage vision transformer (ViT), DEtection TRansformer (DETR), vision and detection transformer (ViDT), faster region based convolutional neural network (Faster R-CNN), inception v3, and ResNet 50 for the same. We examine the adroitness of the leveraged algorithms at new and unseen locations against the challenges of domain generalisation. We demonstrate the effectiveness of the proposed pipeline using the Caltech camera trap (CCT) dataset.

Keywords: Camera traps · Empty frame removal · Wildlife detection · Wildlife species classification · Domain generalisation · DEtection TRansformer (DETR) · Vision transformer (ViT) · Vision and detection transformers (ViDT) · Inception v3

1 Introduction

Apropos the ecological crisis, there is a recent upsurge in research endeavours to conserve wildlife. Camera traps provide wildlife conservation pursuits with continual records of wildlife activity patterns. However, the colossality and obscurity of camera trap data curtail its usage. A primary bottleneck in extracting wildlife insights from camera trap sequence is the overbearing time required to manually extract the required data and the dependence on wildlife experts for the same [20,27]. Surfeit empty frames and preposterous visual challenges such as low resolution, low illumination, deep camouflage, high degree of occlusion, a small region of interest (ROI), and deceiving animal-like background clutter are the major cause of the hefty time consumption in processing

This work is partially supported by the National Mission for Himalayan Studies (NMHS) grant GBPNI/NMHS-2019-20/SG/314.

M. De Marsico et al. (Eds.): ICPRAM 2021/2022, LNCS 13822, pp. 152–174, 2023.
https://doi.org/10.1007/978-3-031-24538-1_8

camera trap sequences [3]. Therefore, in our previous work [1] to speed up the camera trap data processing, we elucidated deep learning based two-stage end-to-end camera trap processing pipeline for empty frame removal and wildlife detection. Inspired by this, we extend our previous work [1], by converting the two-stage pipeline to a three-stage pipeline by adding a third stage: wildlife classification. Therefore, we propose a three-stage end-to-end pipeline for processing camera trap data. In the first stage, we propose to automatically identify and remove empty frames. In this context, empty frames are frames that do not contain any wildlife; bird, or animal. In the second stage, we propose to detect and localise birds and animals using bounding box(es), and in the third stage, we propose to recognise and classify wildlife species.

In the existing literature deep learning based approaches commonly use convolutional neural network (CNN) based algorithms such as faster region-based convolutional neural network (Faster R-CNN) [24], you look only once (YOLO) [23], and single shot detector (SSD) [16] for processing camera trap sequences. But convolution operation has the limitation that the long-range dependencies between semantic concepts are often not deciphered. Though convolutional filters easily capture edges sparse discriminative features escape [31]. As camera trap images are sparse and sporadic in terms of wildlife content it is needed that long-range dependencies are preserved. We need a remedy for alleviating these impediments. Therefore, in our previous work [1] we used transformer-based algorithms DEtection TRansformer (DETR) [4] and vision transformer (ViT) [8] that correlates wildlife presence with the overall image content. In this work, in addition to DETR and ViT, we propose to use another transformer-based deep learning algorithm vision and detection transformer (ViDT). Transformer-based algorithms had an upper hand in the task of empty frame removal and wildlife detection in our previous work [1]. However, in the existing literature, convolution neural networks (CNNs) are eminent for the classification task. Thereby, in this work for wildlife species recognition in addition to DETR, ViT, and ViDT we propose to use inception v3 [30] and ResNet 50 [12].

Camera trap studies are spread across a wide range of camera trap locations. Every biome or natural habitat within the scope of the study is unique due to characteristic flora and fauna, weather and climate patterns, soil, temperature range, day and night duration, and amount of water available. However, deep learning algorithms leveraged for the camera trap processing pipeline cannot be trained on all the locations due to frequent data collection constraints at several locations. The major hindrance to data collection is geographic barriers, adverse weather conditions, and a limited number of camera trap operators. Considering which, in our previous work [1] we critically examined the applicability of the leveraged deep learning based techniques against the challenges of domain generalisability such that pipeline is deemed deployable at new and unseen test locations. In this work, we increase the size of the testbed for unseen locations to encompass a broader range of challenges associated with domain generalisation. Increasing the size of the testbed for unseen locations is particularly required for species classification tasks. The reason is that several species appear different at different locations due to variations in geographic factors.

The approaches available in the existing literature discuss some of the challenges; wildlife detection and species classification etc. in isolation [3,5,33]. Thus, there is a need to address these challenges in a single pipeline. Considering this, we had addressed

(i) empty frame removal and (ii) detection through an end-to-end pipeline in our previous work [1]. In this work, in addition to addressing (i) empty frame removal, (ii) wildlife detection, and (iii) wildlife species classification from camera trap sequences through an end-to-end pipeline, we compare the performance of the proposed pipeline with the pipelines from the existing literature.

Briefly, our contributions in this work are:

- a three-stage end-to-end pipeline for processing camera trap data
- addressing the task of empty frame elimination from camera trap sequences with a competitive performance
- addressing wildlife detection and species classification in camera trap sequences and set a benchmark performance
- critical assessment of leveraged algorithms and each stage of the pipeline against the challenges of domain generalisability
- explicating the suitability of the most competent algorithms via attention maps

The contents of the subsequent sections in this paper are; Sect. 2 reviews the existing literature, Sect. 3 elucidates our proposed pipeline for processing camera trap sequences and proposed approaches to address empty frame removal, wildlife detection, and species classification task, Sect. 4 describes the experiments, results, and inferences for experiments on empty frame removal, wildlife detection, and species classification task and presents the proposed pipeline for processing camera trap sequences, and Sect. 5 provides the conclusion of this work.

2 Related Works

Wildlife detection system finds their application not just in conservation endeavours but also in collision avoidance systems for roads near wildlife sanctuaries [19]. For developing collision avoidance system scale invariant feature transform (SIFT) [18], speeded up robust features (SURF) [2] and continuously adaptive mean shift (CAMShift) algorithm is used. CAMShift is suitable for scenarios where the object to be detected has high contrast against the background [9, 13]. Camouflage is a natural survival and adaptation strategy for wildlife. Therefore, the usage of CAMShift in wildlife detection systems may not be a wise proposition. Some approaches use handcrafted features to detect wildlife [10, 28]. Low-pixel level changes are used to track wildlife. However, these approaches are not suitable when ROI is small [9, 10, 19, 28]. Due to the fixed camera perspective, wild animals captured are often too far, too close or only a part is visible. Thus, the detection system in the wild should be size invariant. Local binary pattern (LBP) [11] and histogram of gradients (HOG) [7] is investigated in [35] for generating features for wildlife detection. LBP [21] was mainly designed for monochromatic and static images. An iterative embedded graph cut (IEC) based method was designed for region proposals in [34]. However, this method is prone to generate false positives due to shadows, waving leaves, and moving clouds.

Ensemble of CNNs namely: AlexNet [14],VGG [26], NiN [15], ResNet [12], and GoogLeNet [29] is used for detecting wildlife from camera trap sequences in [20]. In [3], domain generalisation is explored while using Faster R-CNN and inception v3 for detection and classification. In [25], Faster R-CNN and YOLOv2.0 is used for detection. For wildlife detection and classification, most approaches explore CNN based models. But CNNs have restricted receptive field.

A few approaches elucidate pipelines for processing camera trap sequences [5,33] [3]. In [5], a two-stage pipeline is introduced with the first stage for wildlife detection and the second stage for individual recognition of patterned animals. A one-stage pipeline consisting of species classification is presented in [33]. Some works use a two-stage pipeline with the first stage as wildlife detection and the next stage as species classification [3]. All these approaches manually remove empty frames i.e. the frames captured in camera trap sequences that do not contain any bird or animal part or whole. To remove empty frames every nook and corner of about millions of frames captured in a camera trap sequence is manually examined exhaustively for months and even years. Only a handful of works address the problem of automating empty frame removal [6].

3   Proposed Pipeline to Process Camera Trap Sequences

We propose a three-stage end-to-end pipeline for processing camera trap sequences. The three stages in the proposed pipeline are: (i) empty frame removal, (ii) wildlife detection and localization using bounding box(es), and (iii) wildlife species classification as shown in Fig. 1. In the first stage: empty frame removal, we propose to identify wildlife containing frames and remove the frames in which no wildlife content is captured. The input to stage 1 is camera trap sequences captured at the source and the output is retained wildlife containing frames while empty frames are discarded. In the second stage: wildlife detection, we propose to detect and localize birds or animals with tight-fit bounding boxes. The input to the second stage is the retained wildlife containing frames obtained from stage 1 and the output is the localized birds and animals with bounding box(es). In the third stage: species classification, we propose to recognise and classify the wildlife species. The input for the third stage is the cropped bird or animal images using bounding boxes obtained from stage 2 and the output is the predicted species label. We propose to critically assess the proficiency of each stage in the pipeline against the challenges of domain generalisability.

The comparison of the stages in each camera trap processing pipeline in the existing literature and our proposed pipeline is given in Table 1.

Table 1. Stages in camera trap processing pipeline in existing literature and our proposed pipeline.

Stages	Pipelines			
	Ours	From [5]	From [3]	From [33]
Empty frame removal	✓	×	×	×
Wildlife detection	✓	✓	✓	×
Species classification	✓	✓	✓	✓

Empty Frame Removal: Transformers lack inductive biases: translation equivariance and locality in contrast to CNNs [8]. We hypothesise that translation invariance and

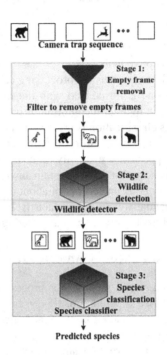

Fig. 1. Proposed pipeline for processing camera trap sequence.

absence of regionally restricted receptive fields in transformers may be suitable for discarding empty frames from camera trap sequences as empty frames do not have any particular object or locality to distinguish them from wildlife containing frames. Empty frames are a collection of a large number of natural scenes of the jungle, grasslands, deserts, river banks, rural dwellings, etc. on a variety of geographic terrains without any bird or animal. Persuaded by this notion, we propose to use transformer-based algorithms ViT, ViDT, and DETR for empty frame removal.

Wildlife Detection: Animals in the wild are often occluded leading to a discontinuity in animal image pixels. Despite discontinuity in animal image pixels, it is needed that the distant but related animal features should be interweaved to correctly detect and localise the animal in a tight-fit bounding box. In addition to discontinuity in animal image pixels due to occlusions, the camera trap data is crammed with nighttime images, negligible tonal gradient, variations in animal sizes and poses, and camouflage. In the existing literature deep learning based approaches commonly use convolutional neural network (CNN) based algorithms such as faster region-based convolutional neural network (Faster R-CNN) [24], you look only once (YOLO) [23], and single shot detector (SSD) [16] for processing camera trap sequences. But convolution operation has the limitation that the long-range dependencies between semantic concepts are often not deciphered. Though convolutional filters easily capture edges sparse discriminative features escape [31]. For extracting bionomical statistics from camera trap sequences, alternatives alleviating these encumbrances are needed. Therefore, we propose to experiment with DEtection TRansformer (DETR) [4]; a recent object detection algorithm that computes relationships between wildlife pixels with all the image pixels. DETR comprises transformer attention-based encoder-decoder stacks for performing global reasoning. The obscured animal parts can be deciphered by the means of global reasoning. Low illumination, negligible tonal gradient and occlusion being predominant, transformers should have an upper hand. Building upon the notion that the transformer-based models are suitable for wildlife detection in camera trap sequences, we choose to explore another transformer based model ViDT. ViDT generates hierarchical features capturing context semantics across different scales and sizes. The cross-attention between image patches and region proposals teases out patterns between related heterogeneous concepts. We envisage that capturing rich feature semantics across different scales and sizes and cross-attention will help in the wildlife detection task. The reason is that the birds and animals captured in camera trap sequences are inordinately variable in sizes. Therefore, we propose to use DETR and ViDT for wildlife detection.

We hypothesise that transformer-based algorithms in comparison to purely convolution-based algorithms will prove to be more suitable for handling the adverse challenges of data from camera traps, especially for addressing the empty frame removal and wildlife detection tasks. Thus, for comparison and to validate our hypothesis we choose Faster R-CNN.

Species Classification: Transformer attention-based classification models outperform CNN based classification model only under very specific preconditions [8,22]. Even under specific conditions, the percentage improvement in performance given by transformer-based classification model in comparison to CNNs is marginal [8]. Hence, transformers are yet to prove their competence over CNNs in an image classification task. Therefore, in addition to ViT, DETR, and ViDT we rope in a state-of-the-art CNN-based image classification model inception v3 [30]. Inception v3 is built by stacking several inception modules composed of several factorised convolution blocks [30]. Factorised convolution in comparison to larger convolutions is more suitable for tasks that require the model to decide based on minuscule features. A larger convolution filter quickly diminishes the size of the input feature maps causing minuscule and important discriminative features to be lost. On the contrary, factorised convolution diminishes the feature maps gradually through multiple hierarchical and parallel smaller convolution operations. It is felt that for species recognition and classification from camera trap sequences factorised convolution will be helpful. This is because in camera sequences the animal captured are far, small in comparison to large swaths of vegetation and sky, and frequently only a part of the animal is captured. Thus, we propose to use inception v3 for species recognition and classification task from camera trap sequences. In addition to inception v3, we propose to use another state-of-the-art CNN classification model ResNet 50 for recognising and classifying species from camera trap sequences. The building block of ResNet 50 is a residual block. A skip connection in a residual block concatenates the input of a previous layer with the output of the current layer. The residual blocks ensure that minuscule context semantics are preserved throughout using heterogeneous feature map concatenation. Thus, the residual blocks will be particularly helpful in species recognition and classification from camera trap sequences.

4   Experimental Studies for Processing Camera Trap Sequences

This section discusses the experimental studies on processing camera trap sequences. Our experimental study is divided into; (1) studies on empty frame removal, (2) studies on wildlife detection and localization, and (3) studies on species classification. All these studies are performed on Caltech camera traps (CCT) dataset [3].

4.1   Experimental Studies on Empty Frame Removal

We first discuss the Caltech camera traps (CCT) dataset and data split for addressing empty frame removal.

Caltech Camera Traps (CCT) Dataset and Data Split for Addressing Empty Frame Removal: Caltech camera traps (CCT) dataset is a camera trap sequence data that was collected unobtrusively from Southwestern United States across 140 locations. It is a sequence of 243,100 frames. The wildlife recorded in this dataset is badger, bat, birds, bobcat, cat, coyote, cow, deer, dog, fox, insect, lizard, mountain lion, opossum, pig, rabbit, raccoon, rodent, skunk, and squirrel. Alike most camera trap sequences more than 70% frames are empty. Therefore, the distribution of frames is biased toward frames that are empty as illustrated in Fig. 2. Although on average more than 70% frames are empty, from Fig. 2 it is seen that the percentage of empty frames per location greatly varies between 10%-97%. Here, an wildlife/animal frame means an image with an animal or bird present and an empty frame means an image with no animals or birds. Before experimentation, we select equal numbers of wildlife containing frames and empty frames for the training data from each location. This is done to prevent the leveraged algorithms from developing a bias towards the more commonly occurring class; empty frames. The data split is given in Table 2. The training data comprises camera trap sequences from 20 random locations. A total of $8,028$ images having equal numbers of wildlife containing frames and empty frames is extracted, from which 70% is used for training and 30% is used for testing. Thus, $5,638$ training and $2,390$ testing images are used. We create another set for testing from the remaining 120 locations that are not seen by the algorithm at the time of training. By additionally testing our models on unseen locations we analyze the domain generalisability of the models. Therefore, all in all, we create two test sets; (1) 'cis': from 20 locations seen by the algorithm at the time of training and (2) 'trans': from 120 locations not seen by the algorithm at the time of training. Therefore, the 'trans' set consists of frames from 120 locations accounting for $105,745$ empty frames and $35,498$ wildlife frames with background characteristics different from 'cis' locations.

Table 2. Data splits for empty frame removal.

	Number of locations	Wildlife frames	Empty frames	Total
Train: 'cis'	20	2819	2819	5638
Test: 'cis'	20	1195	1195	2390
Test: 'trans'	120	25,000	105, 745	141, 243

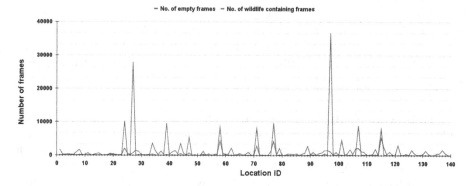

Fig. 2. Number of wildlife containing frames and empty frames vs. locations in the CCT dataset.

Experimental Setup and Results for Empty Frame Identification and Removal:
ViT, DETR, ViDT, and Faster R-CNN is used for segregating empty frames from
wildlife frames. Empty frame identification is posed as a classification problem with
two classes: (i) empty frames and (ii) frames containing wildlife. We use percentage
accuracy (% of images correctly classified into an empty or wildlife frame) as the metric for evaluating performance. The models are trained on 'cis' locations and tested on
both 'cis' and 'trans' location sets.

The ViT model is trained using AdamW [17]. The weight decay parameter is set
to 0.0001 and the initial learning rate is 0.001. Random rotation, horizontal flip and
zoom are the data augmentation techniques used in conjunction with ViT experiments.
DETR and ViDT models are finetuned with AdamW [17] with 0.0001 learning rate.
Random horizontal flips is employed for data augmentation in experiments with DETR
and ViDT. When training images are passed to the trained model, it is observed that
all wildlife frames are recognised with a confidence score of 0.95. Therefore, 0.95 is
chosen as the confidence threshold for identifying and sieving out empty frames. While
testing, a frame under scrutiny is deemed empty if the confidence score for wildlife presence is below 0.95. Pretrained Faster R-CNN with ResNet-101 backbone is availed from
Detectron2 codebase [32]. For experiments with Faster R-CNN the absence of wildlife
is marked in the empty frames by using $[0, 0, 0, 0]$ co-ordinate values for bounding box
annotations; to represent null dimensions and area. A null tensor for both confidence
score and bounding box is expected as output for correct identification of empty frame.
Stochastic gradient descent (SGD) with momentum set at 0.9 is used. The initial learning rate 0.001 is decayed by a factor of 0.05 after every 1000 epochs.

The performance of different deep learning techniques: ViT, DETR, ViDT, and
Faster R-CNN on empty frame identification and removal is given in Table 3. Among
these four algorithms, the best accuracy (87.28%) for 'cis' locations is given by ViT.
The best overall performance on 'trans' locations 67.83% is set by DETR. DETR, ViDT
and Faster R-CNN for wildlife containing frames bestow an accuracy of greater than
90% but for an empty frame, the accuracy is 61.51%, 59.51 and 40.59% respectively.
These models detect birds or animals with high precision but a significant number of
empty frames are misclassified as wildlife containing frames. Thus the objective of

identifying and discarding empty frames will not be met. Especially in the case of Faster R-CNN, there are many false positives as more than 66% of empty frames are being recognised as wildlife frames. At both 'cis' and 'trans' locations we see that ViT has the highest accuracy in correctly classifying empty frames. The primary objective of empty frame removal is to identify empty frames correctly. Therefore, we choose ViT for empty frame removal.

Domain Generalisability: It is observed from Table 3 that ViT gives 87.28% accuracy on 'cis' and 64.28% accuracy on 'trans'. This provides evidence of the existence of a generalisation gap between seen and unseen domains. At 'cis' locations, the precision for correctly identifying empty frames is higher and on the contrary, for 'trans' locations the precision for correctly classifying wildlife frames is higher. The background at a particular location for empty frames and wildlife containing frames is the same. Hence, for 'cis' locations the algorithm develops a bias for learning empty frame characteristics. On the flip side, for 'trans' locations the same wildlife species as in 'cis' are seen but the background differs leading to higher precision for classifying wildlife frames. Faster R-CNN gives an accuracy of 65.35% at 'cis' locations and accuracy of 45.24% at 'trans' locations, and hence there exists a performance gap across seen and unseen locations. ViDT gives an accuracy of 78.90% on 'cis' locations and an accuracy of 65.08% on 'trans' locations. The precision obtained in classifying wildlife frames in both 'cis' and 'trans' is fairly higher than in classifying empty frames. We deduce that the generalisation gap is fairly less in ViDT due to the hierarchical cross-attention mechanism. DETR gives an accuracy of 80% on 'cis' locations and an accuracy of 67.83% on 'trans' locations. It is inferred that DETR shows least generalisation gap due to transformer attention-based encoder-decoder self and cross attention.

Table 3. Accuracy in % for empty frame identification from ViT, DETR, ViDT and Faster R-CNN.

Model	'cis'			'trans'		
	Wildlife	Empty	Total	Wildlife	Empty	Total
ViT	84.60	89.96	87.28	69.29	63.26	64.28
DETR	98.49	61.51	80.00	90.71	62.43	67.83
ViDT	98.29	59.51	78.90	88.76	59.43	65.08
Faster R-CNN	98.74	31.96	65.35	95.06	33.47	45.24

Supposedly, we extend our algorithms for detecting rare species. We would retain any frame having even the slightest hint of an uncommon species as they are extremely difficult to capture in camera trap sequences. Therefore, we would slacken discarding empty frames if the price is losing frames containing rare species. DETR, ViDT, and Faster R-CNN have nearly the same performance while classifying wildlife frames (98.49%, 98.29%, and 98.74% respectively) on 'cis' locations. However, DETR outperforms Faster R-CNN approximately by 30% and ViDT by 2% on 'cis' locations. On 'trans' locations DETR has nearly the same performance as ViT ($\approx$ 63%) for empty frame removal, but it is the most suitable model for detecting infrequently encountered species. This is so because DETR recognises 90.71% wildlife frames correctly and only 9.29% of wildlife frames get misclassified as empty frames. In comparison, Faster R-CNN (95.06%) gives greater accuracy than DETR in detecting wildlife frames but the number of empty frames it is detecting correctly is 30% less for 'trans' locations, totally defeating the purpose of empty frame removal. The presence of surfeit empty frames may be counterproductive and may obfuscate the process of recognising rare species in Faster R-CNN.

Generally, camera traps record colossal amounts of data consisting of more than 70% surfeit empty frames. Thus, the sweet spot between detecting wildlife species and eliminating empty frames can be achieved by using ViT. To further affirm ViTs prowess on the empty frame removal task, we visually scrutinize the performance of ViT on 1000 randomly selected frames from 'cis' and 'trans' location sets each by plotting self-attention maps. The plotted self-attention maps help us to verify if the attention is indeed focusing itself on birds and animals present in wildlife containing frames. Some self-attention maps plotted for visual scrutiny are shown in Fig. 3. From Fig. 3 (a) and (b) we see that the attention is solely focused on the animal in wildlife containing frames. From Fig. 3 (c) and (d) we observe that the attention is not localized but dispersed in empty frames. Approximately 82% and 73% of the attention maps for wildlife containing frames from 'cis' and 'trans' respectively have attention rightly focusing itself on the bird or animal present in the frame irrespective of pose variations, poor illumination, masquerading camouflage, low contrast and occlusion. Further, we observe that for approximately 90% of the empty frames at 'cis' locations the attention score is negligible, and for 61% of the empty frames at 'trans' location the attention is either scattered or negligible. Therefore, through visualization of attention maps, it is seen that indeed ViT accurately focuses its attention on the birds or animals present and is consignable for the camera trap processing pipeline.

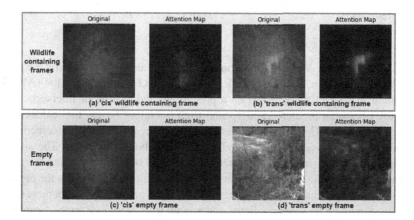

Fig. 3. Attention map from ViT: (a) 'cis' and (b) 'trans': The attention is rightly focused on the animal in frames containing wildlife. (c) 'cis' and (d) 'trans': The attention is not focused but negligible or dispersed in empty frames.

The next section presents results on detecting (localizing) wildlife.

4.2   Experimental Studies for Wildlife Detection

The dataset used and data split for the wildlife detection task is discussed first. Then we discuss the results on wildlife detection and localization.

CCT Data for Wildlife Detection and Localization: We train our model for wildlife detection and localization task on 20 randomly chosen 'cis' locations same as in the empty frame removal task. We opt to test the applicability of our models on both 'cis' and 'trans' locations to critically assess the prowess of our models in the light of domain generalisation. The data split for the wildlife detection task is given in Table 4. The number of training images used for the wildlife detection task is 13,553 with 14,071 bounding box annotations. The validation set has 3,484 wildlife images with 3,582 bounding box annotations. The 'cis' test set has 15,827 images with 16,395 bounding box annotations. The 'trans' set consists of 120 unseen locations with 25,000 test images and 25,895 bounding box annotations. The 'trans' set of locations is contrived to be completely unseen by the model, hence we do not extract validation data from 'trans' locations.

Table 4. Data splits for wildlife detection task.

Data	Number of locations	Number of frames	Number of bounding box annotations
Train: 'cis'	20	13,553	14,071
Validation: 'cis'	20	3,484	3,582
Test: 'cis'	20	15,827	16,395
Test: 'trans'	120	25,000	25,895

Evaluation Metric for Wildlife Detection: The metric used for evaluating the performance of algorithms for wildlife detection task is COCO average precision (AP). The computation of COCO AP is based on a preliminary calculation of Intersection over Union (IoU) between predicted and ground truth bounding boxes. IoU calculation is illustrated in Fig. 4. IoU is calculated as the ratio between the area of intersection and the area of union between predicted and ground truth bounding boxes. The area that is common overlaps or intersects between the predicted bounding box and the ground truth bounding box is the area of intersection. The resulting area if we fuse the predicted and ground truth bounding boxes together is the area of union. Figure 5 depicts sample IoU scores computed between predicted and ground truth bounding boxes. From Fig. 5 (a) we see that an IoU score of 0.9 is nearly a perfect overlap between the predicted and ground truth bounding boxes. Hence, an IoU score of 0.9 and above is considered near-perfect detection. From Fig. 5 (c) we can observe that an IoU score of 0.33 between the predicted and ground truth boxes is greater than the 50% area overlap. It is interesting to note from Fig. 5 (d) that a seemingly low IoU score 0.14 is an area-wise overlap of nearly 25% between predicted and ground truth boxes. The bottom line is that the IoU score being a ratio between the area of intersection and area of union should not be directly perceived as a percentage of area overlap. Generally, the percentage of area overlap is greater than the absolute IoU score. We observe that IoU is indeed a strict metric.

After the preliminary calculation of IoU scores between predicted and ground truth bounding boxes, the next step for the computation of COCO AP is to demarcate the predictions into true positives (TP), false positives (FP), false negatives (FN), and true negatives. In the detection paradigm, a detection is considered a true positive (TP) if the IoU score between the predicted and ground truth box is above a threshold, otherwise, it is considered a false positive. In Fig. 6 (a), the IoU score between the predicted and ground truth bounding box is 0.53. If the IoU threshold is set at 0.5, then Fig. 6 (a) is a true positive (TP). On the contrary, if the IoU threshold is set at 0.75, then it is a false positive (FP). The case where there is no detection for a frame with an animal is called false negative (FN) as shown in Fig. 6 (c). A true negative (TN) is an empty frame that does not contain any animal. After demarcating TP, FP, and FN, precision ($\frac{TP}{TP+FP}$) and recall ($\frac{TP}{TP+FN}$) are calculated for a decided IoU threshold.

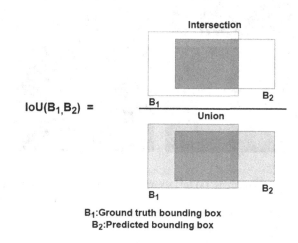

Fig. 4. Intersection over union (IoU) calculation.

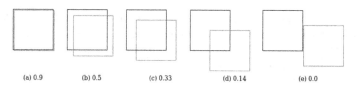

Fig. 5. Sample IoU scores (boxes in green are ground truth and boxes in blue are ground truth). (Color figure online)

Thereby, COCO AP 0.5–0.95 is calculated as the average precision over multiple IoU thresholds (10 IoU thresholds) ranging from 0.5 to 0.95 with a step size of 0.05. Mathematically, the average precision (AP) is defined as the average of precision values at 101 equally spaced recall levels [0, 0.01 ..., 1] at a step size of 0.01 as given below:

$$AP = \frac{1}{101} \sum_{r \in \{0.0, 0.01, ..., 1\}} p_{\text{inter-p}}(r) \tag{1}$$

The interpolated precision, $p_{\text{interp}}(r)$ in the above equation, is calculated at each recall level, r, by taking the maximum precision measured for that r :

$$p_{\text{interp}}(r) = \max_{\tilde{r}:\tilde{r} \geq r} p(\tilde{r}) \tag{2}$$

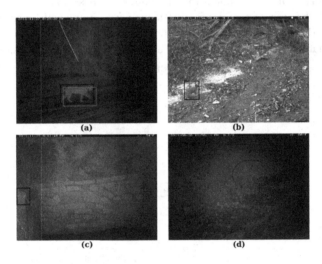

Fig. 6. Sample bounding predictions. (a) IoU score is 0.53; a true positive (TP) if the IoU threshold is 0.5 and a false positive (FP) if the IoU threshold is 0.75. (b) IoU score is 0.37; a false positive (FP) because the IoU threshold is generally set at 0.5 or above. (c) No detection output; false negative. (d) True negative as the frame is empty and has no animal in it. (The boxes in red are ground truth and in blue are predicted). (Color figure online)

In Eq. 2 $p(\tilde{r})$ is the measured precision at recall $\tilde{r}$. COCO AP 0.5 and COCO AP 0.75 is calculated on similar lines as that of COCO AP 0.5–0.95. The difference is that instead of using multiple thresholds, 0.5 and 0.75 are taken as IoU thresholds for COCO AP 0.5 and COCO AP 0.75 respectively.

Experimental Results for Wildlife Detection and Localization Task: The performance of DETR, ViDT and Faster R-CNN on wildlife detection is given in Table 5. It can be seen that DETR and ViDT have less generalisation gap and have an upper hand over Faster R-CNN in the wildlife detection task. It is deduced that computation of pixel-wise relationship by the self and cross attention mechanism is the reason for the superior performance of DETR and ViDT. DETR has marginally higher performance than ViDT. Thus, we employ DETR for further scrutiny.

Table 5. Wildlife detection results in AP (COCO AP 0.5–0.95).

Test data	Faster R-CNN	DETR	ViDT
'cis'	54.8	56.8	56.0
'trans'	52.9	55.2	55.1

Domain Generalisability: In the case of wildlife detection, we observe from Table 5 that the generalisation gap is negligible across 'cis' and 'trans' locations for any of the three algorithms.

Sequence Analysis of Camera Trap Images with Results from DETR: Whenever the camera trap sensor receives a trigger to capture images, a varying length of sequence consisting of 1–5 frames is recorded [3]. It is assumed that at least one frame will record the bird or animal. Thus, we exploit frame sequence information as in [3] in two ways:

1. **Most Confident:** Wildlife instance in a sequence of frames is considered correctly detected if the frame associated with the highest detection score in the sequence has an IoU more than 0.5.
2. **Oracle:** Wildlife instance in a sequence of frames is considered accurately detected if any frame has an IoU more than 0.5.

For sequence analysis frames with multiple wildlife instances are not considered. The results for sequence analysis are given in Table 6. Without sequence information an average precision (COCO AP 0.5) of 90.1 on 'cis' and 89.2 using on 'trans'. With sequence information 'most confident' and 'oracle' we wield an average precision (COCO AP 0.5) of 93.0 and 96 respectively on 'cis'. With sequence information 'most confident' and 'oracle' we obtain average precision (COCO AP 0.5) of 91.4 and 94 using respectively on 'trans'. Multiple frames in a sequence are shot to capture wildlife instances accurately at least in one frame. Therefore, exploiting sequence information for detection evaluation is appropriate. 94 average precision (COCO AP 0.5) points on 'trans' seem to be up to the mark against the dire challenges in camera trap sequences; motion blur, camouflage, only a part of bird or animal captured, low illumination, nearly flat tonal gradient and occlusion.

Table 6. Sequence analysis of camera trap data using DETR.

	'cis'			'trans'		
	No sequence information	Most Confident	Oracle	No sequence information	Most Confident	Oracle
AP 0.5–0.95	56.9	62.9	69.4	55.4	61.2	68.5
AP 0.5	90.1	93.0	96	89.2	91.4	94

For further scrutinizing the performance of DETR on camera trap sequences 1000 images are randomly selected. Some images used for visual scrutiny are shown in Fig. 7. Near perfect detection is observed in 93.1% of the daytime images (Fig. 7 (a)) and in 81.39% of low-light images (Fig. 7 (b) and (c)). Despite the lack of luminance and colour gradient in a fair share of images, DETR accurately locates birds or animals in 85.77% images. Most poor localisations arises due to deceiving animal-like background clutter (Fig. 7 (d)), due to extreme low illumination (Fig. 7 (e)), and due to deep camouflage (Fig. 7 (f)). Certainly, in many such cases, it is not evident if the animal is present or not.

The DETR encoder self-attention using four reference points is shown in Fig. 8. The self attention visualised for a reference point illustrates the computed relationship between the reference points and the remaining points. This visualisation gives insights into how accurately the model learns to knit context semantics. In Fig. 8 it is seen that for all reference points belonging to the deer ((a), (b) and (c)), the model assigns maximum weight to the point belonging to the deer. The self-attention matrix visualised for three spatially apart reference points belonging to the deer is approximately the same. Thus, it is deduced that the model correctly knits related context semantics. It is also observed that in each self-attention map (Fig. 8 (a), (b) and (c)) the background pixels are weighted less, and hence the model learns to filter out background clutter. Further, for a background reference point (d), the overall background pixels are assigned higher weight in comparison to animal pixels. This reaffirms that the DETR encoder weaves long-range dependencies. The ability to have a global perspective beyond the local restrictive field of a convolution filter is the primary reason for the better performance of DETR on camera trap sequences.

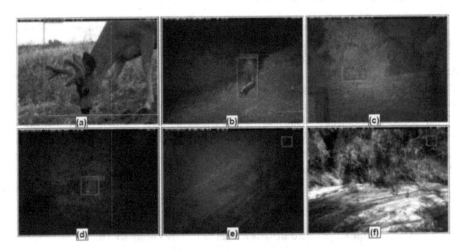

Fig. 7. DETR detection output on 'trans' locations. (a) IoU equal to 0.99, (b) IoU equal to 0.79, (c) IoU equal to 0.63, (d) False positive; IoU equal to 0.39, (e) False negative detection, and (f) False positive detection (The bounding boxes in red are ground truth and boxes in blue are predicted). (Color figure online)

Fig. 8. Visualization of DETR encoder self-attention weights for a 'trans' image; (a), (b) and (c): self attention map for reference point belonging to the animal. (d): self attention map for reference point belonging to background.

4.3   Experimental Studies on Species Recognition from Camera Trap Sequences

The dataset and data split for species classification task from camera trap sequences are discussed beneath.

CCT Data for Species Recognition from Camera Trap Sequences: We use the CCT dataset for species recognition from camera trap sequences. The wildlife category labels identified in the dataset are: $badger$, bat, $birds$, $bobcat$, cat, $coyote$, cow, $deer$, dog, fox, $insect$, $lizard$, $mountain$ $lion$, $opossum$, pig, $rabbit$, $raccoon$, $rodent$, $skunk$, and $squirrel$. We split the data in congruence to empty frame removal and wildlife detection experiments. The data split is shown in Table 7. $13, 553$ images with species labels from 20 'cis' locations are apportioned for training. The 'cis' test set has $15, 827$ images and 'cis' validation set has $3, 484$ images. The 'trans' has $25, 0000$ images for testing.

Table 7. Data splits for species classification task.

Data	Number of locations	Number of images
Train: 'cis'	20	13553
Validation: 'cis'	20	3484
Test: 'cis'	20	15827
Train: 'trans'	120	25000

Experimental Results for Species Classification Task from Camera Trap Sequences: DETR, ViDT, Faster R-CNN, ViT, inception v3, and ResNet 50 are used for our experiments. For species classification, the input to the algorithms is the output of the second stage from the proposed pipeline for processing camera trap sequences. We use percentage accuracy (% of images correctly classified into the respective species category) as the evaluation metric. DETR and ViDT models are finetuned with AdamW with a learning rate of 0.0001. Random horizontal flip is used as the data augmentation technique. Congruous with our experiments on empty frame removal and wildlife detection this choice of hyperparameters is found to be most suitable. Pretrained Faster R-CNN with ResNet-101 backbone is availed from Detectron2 codebase [32]. Stochastic gradient descent (SGD) with momentum set at 0.9 is used. The initial learning rate 0.001 is decayed by a factor of 0.05 after every 1000 epochs. Inception v3 and ResNet 50 are trained using Adam with a learning rate of 0.0001. ViT is trained using AdamW with a learning rate of 0.001.

The experimental results of species classification from the leveraged deep learning techniques DETR, ViDT, Faster R-CNN, ViT, inception v3 and ResNet 50 are given in Table 8. ViDT gives 90.77% accuracy on 'cis' locations and 72.12% accuracy on 'trans' locations. DETR gives 86.85% accuracy on 'cis' locations and 61.90% accuracy on 'trans' locations. ViDT and DETR algorithms are consignable considering the adverse challenges in species identification from camera trap sequences. The merit of ViDT and DETR can be attributed to the coalescing of context semantics by attention-based global reasoning and pixel-wise self-attention. Faster R-CNN gives 74.2% accuracy on 'cis' locations and 52.08% accuracy on 'trans' locations. Congruence to our observation in empty frame removal and wildlife detection experiments, the performance of Faster R-CNN is subordinate to DETR and ViDT in camera trap sequence processing. From Table 8 it seen that 91.75%, 74.58% and 84.93% accuracy is given by inception v3, ViT and ResNet 50 respectively on 'cis' locations and 72.83%, 55.18% and 68.42% is given by inception v3, ViT and ResNet 50 on respectively 'trans' locations. We see that the performance of ViT is subordinate to inception v3 and ResNet 50. In agreement with the existing literature, we see that the transformer attention-based classification models are yet to prove their dominance over CNN-based classification models in image classification tasks. The overall highest accuracy is given by inception v3. The success of inception v3 is accredited to the inception modules comprising the factorised convolution operation. Hence, experimentally it is established that the factorised convolution is indeed helpful in capturing context semantics and features across various scales and hierarchies in the case of camera trap sequences.

Table 8. Species classification results from camera trap sequences (accuracy in %).

Test location	DETR	ViDT	Faster R-CNN	ViT	Inception v3	ResNet 50
'cis'	86.85	90.77	74.2	74.58	91.75	84.93
'trans'	61.90	72.12	52.08	55.18	72.83	68.42

Domain Generalisability: Consistent across all algorithms, we observe from Table 8 that the performance on 'cis' locations is better than on 'trans' locations. This observation directly establishes a link between the difference in image distribution and characteristics in 'cis' and 'trans' location sets. Upon visual scrutiny, it is observed that the 'trans' locations are comparatively more challenging due to the larger degree of motion blur observed in nighttime images causing the animals in 'trans' to appear less prominent. Another major factor causing the algorithms to work less proficiently on 'trans' locations is high intra-class variance, especially amongst *bird, cat, dog, deer* and *rabbit*. At 'trans' locations different kinds of birds are present, different breeds of *dog* and *cat* are present, sexual dimorphism is observed in deer, and the fur and coat of *rabbit* vary in colour. In addition, to intra-class variance high inter-class similarity is also observed. For example; *coyote* and *fox* appear very similar and are difficult to tell apart, especially in images with only a part of *coyote* or *fox* captured. Despite the significant observable difference in flora and fauna at 'trans' locations inception v3 and ViDT generalise well to unseen locations.

Experiment to Verify the Need for Detection Prior Species Classification: In the above experiment set following the proposed approach for processing camera trap sequences, we feed camera trap sequences into the classification models after empty frame removal and wildlife localization and detection. To check the significance of detection and localization before species classification we pass camera trap sequences to ViT, inception v3 and ResNet 50 directly after the empty frame removal stage. The results for species classification from full images are given in Table 9. We observe that the performance of inception v3, ResNet 50 and ViT significantly drops when fed with full images from camera trap sequences. The presence of background confounds the algorithms. Therefore, we conclude that for species recognition from camera trap sequences it is indeed required that after removing empty frames, birds and animals should be localized and detected before feeding to classification algorithms.

Table 9. Species classification results from camera trap sequences on full images (accuracy in %).

Test location	Inception v3	ResNet 50	ViT
'cis'	51.64	54.52	48.49
'trans'	35.39	37.58	31.54

We culminate the experimental analyses on species classification from camera sequences and conclude that inception v3 is the most suitable choice amongst all the leveraged algorithms.

Discussion on the Proposed Pipeline: Persuaded by our results, we propose a three-stage end-to-end pipeline for empty frame removal, wildlife detection and species identification in camera trap sequences. The proposed pipeline is shown in Fig. 1. The proposed pipeline first filters out the empty frames and then locates and finally recognises and classifies the species of the bird or animal present. We use ViT for empty frame removal, DETR for detection and inception v3 for the classification task. The proposed pipeline is applied to the entire data and 85.42% of the total empty frames are discarded. Approximately 76 GB of total 102 GB of the camera trap sequence data used for testing contains empty frames. By removing 85.42% of empty frames, roughly 65 GB of the space used in futile for storing empty frames is freed. In the next stage, 93.41% of the wildlife is detected with IoU greater than 0.5 and 98.56% of the wildlife is detected with IoU greater than 0.3. In the third stage, 90% of birds and animals are classified correctly. The results from the pipeline are shown in Fig. 9. We see in Fig. 9 (a) that despite low illumination the *raccoon* is correctly detected and classified. In Fig. 9 (b) it is seen that even if the part of *coyote* is captured, it is detected and classified correctly. In Fig. 9 (c) it is seen that the *cat* is detected correctly but incorrectly classified as *badger*. In Fig. 9 (c), it is indeed difficult to see the animal due to occlusion and extremely low light. In Fig. 9 (d), the rabbit is detected with IoU less than 0.5 but is correctly classified. We maintain our pipeline to be of three stages considering the dire challenges of camera trap sequences wherein in many cases it is likely the detection is accurate but the classification is not and vice versa. Therefore, a three stage pipeline allows us to leverage at least one correct output.

We compare the performance of the proposed camera trap processing pipeline with the camera trap processing pipelines in the existing literature. The pipeline in [3] processes camera trap sequences (containing only wildlife containing frames) using Faster R-CNN for wildlife detection and Inception v3 for species classification on images cropped using ground truth bounding box annotation. To replicate the pipeline in [3], we input wildlife containing frames into Faster R-CNN for wildlife detection and input image cropped using bounding boxes obtained from Faster R-CNN into Inception v3. We ablate the empty frame removal block in our pipeline and directly feed in camera trap sequences with only wildlife containing frames for comparison with [3]. The comparison result is shown in Table 10. Upon comparison, we see that our algorithms out-

Fig. 9. Results from the proposed pipeline for processing camera trap sequences. (a) *raccoon* correctly detected with IoU 0.9 and correctly classified, (b) *coyote* correctly detected with IoU 0.97 and correctly classified, (c) *cat* detected correctly with IoU 0.79 but incorrectly classified as *badger* (d) *rabbit* detected with IoU less than 0.5 but correctly classified.

perform at both stages. To replicate the pipeline given in [33], we feed in only wildlife containing frames and use DenseNet-161 used for species classification as in [33]. For comparison with the pipeline given in [33], we ablate empty frame removal and wildlife detection blocks and directly feed in camera trap sequences with only wildlife containing frames with cropping. We see from Table 11 that our proposed pipeline outperforms the pipeline given in [33]. The pipeline given in [5] is used for automatic detection and individual recognition in patterned animal species. Therefore, the scope of work in [5] is limited to individual identification and is different from our objectives. Hence, it is not fair to compare the pipeline given in [5] with our pipeline.

Table 10. Comparison of our proposed pipeline with pipeline given in [3] (detection performance in COCO AP 0.5–0.95 and classification accuracy in %).

Stage	Pipelines tested on 'cis'		Pipelines tested on 'trans'	
	Ours	From [3]	Ours	From [3]
Wildlife detection	56.8	54.8	55.2	52.9
Species classification	91.75	83.49	72.83	66.13

Table 11. Comparison of our proposed pipeline with pipeline given in [33] (classification accuracy in %).

Stage	Pipelines tested on 'cis'		Pipelines tested on 'trans'	
	Ours	From [33]	Ours	From [33]
Species classification	91.75	46.1	72.83	31.14

5   Conclusion

We propose a three-stage pipeline for processing camera trap sequences. We envisage that the pipeline for processing camera trap sequences will reduce the time taken to extract valuable statistics. Thus, enabling camera trap sequences to be used for large-scale comprehensive environmental studies. Alone, empty frame removal through deep learning techniques can fasten the camera trap processing by a few months. By discarding empty frames at regular intervals approximately 70% of the total storage space required to store camera trap sequences can be saved. Most camera trap data repositories are untapped due to their colossal size. Unless the camera trap data is used to obtain ecological insights and subsequent environmental policy and decision-making, the deployment of camera traps across locations is futile. Therefore, by addressing wildlife detection and species classification through deep learning techniques we speed up the extraction of ecological statistics from camera trap sequences and render camera trap sequences serviceable. Despite the preposterous challenges: low illumination, negligible colour tones, extreme occlusion, masquerading camouflage, inexact perspective, and small ROI; the proposed approaches give a consignable performance. The visually imperceptible wildlife presence in camera traps can be located and identified decreasing the burden on wildlife experts. Further, we obtained competent performance while critically assessing the adroitness of the proposed approaches in empty frame removal, wildlife detection, and species classification across new and unseen locations. Therefore, the camera trap processing pipeline can be used directly without retraining the leveraged algorithms in new and unseen locations. In future, we plan to make the proposed pipeline even more robust against domain generalisation challenges by incorporating an additional functional block for open-set recognition. Further, as the species distribution in nature is heavily imbalanced and non-uniform, we plan to improve the performance of the leveraged algorithms on less represented classes through few-shot recognition.

References

1. Banerjee, A., Dinesh, D.A., Bhavsar, A.: Sieving camera trap sequences in the wild. In: ICPRAM, pp. 470–479 (2022)
2. Bay, H., Tuytelaars, T., Van Gool, L.: SURF: speeded up robust features. In: Leonardis, A., Bischof, H., Pinz, A. (eds.) ECCV 2006. LNCS, vol. 3951, pp. 404–417. Springer, Heidelberg (2006). https://doi.org/10.1007/11744023_32
3. Beery, S., Van Horn, G., Perona, P.: Recognition in terra incognita. In: Ferrari, V., Hebert, M., Sminchisescu, C., Weiss, Y. (eds.) ECCV 2018. LNCS, vol. 11220, pp. 472–489. Springer, Cham (2018). https://doi.org/10.1007/978-3-030-01270-0_28
4. Carion, N., Massa, F., Synnaeve, G., Usunier, N., Kirillov, A., Zagoruyko, S.: End-to-end object detection with transformers. In: Vedaldi, A., Bischof, H., Brox, T., Frahm, J.-M. (eds.) ECCV 2020. LNCS, vol. 12346, pp. 213–229. Springer, Cham (2020). https://doi.org/10.1007/978-3-030-58452-8_13
5. Cheema, G.S., Anand, S.: Automatic detection and recognition of individuals in patterned species. In: Altun, Y., et al. (eds.) ECML PKDD 2017. LNCS (LNAI), vol. 10536, pp. 27–38. Springer, Cham (2017). https://doi.org/10.1007/978-3-319-71273-4_3
6. Cunha, F., dos Santos, E.M., Barreto, R., Colonna, J.G.: Filtering empty camera trap images in embedded systems. In: Proceedings of the IEEE CVF Conference on Computer Vision and Pattern Recognition (CVPR) Workshops, pp. 2438–2446 (2021)
7. Dalal, N., Triggs, B.: Histograms of oriented gradients for human detection. In: Proceedings of 2005 IEEE computer society conference on Computer Vision and Pattern Recognition (CVPR'05), vol. 1, pp. 886–893. IEEE (2005)
8. Dosovitskiy, A., et al.: An image is worth 16x16 words: Transformers for image recognition at scale. arXiv preprint arXiv:2010.11929 (2020)
9. Emami, E., Fathy, M.: Object tracking using improved CAMshift algorithm combined with motion segmentation. In: Proceedings of the 7th Machine Vision and Image Processing (MVIP), 2011 Iranian, pp. 1–4 (2011)
10. Figueroa, K., Camarena-Ibarrola, A., García, J., Villela, H.T.: Fast automatic detection of wildlife in images from trap cameras. In: Bayro-Corrochano, E., Hancock, E. (eds.) CIARP 2014. LNCS, vol. 8827, pp. 940–947. Springer, Cham (2014). https://doi.org/10.1007/978-3-319-12568-8_114
11. Guo, Z., Zhang, L., Zhang, D.: A completed modeling of local binary pattern operator for texture classification. IEEE Trans. Image Process. 19(6), 1657–1663 (2010)

12. He, K., Zhang, X., Ren, S., Sun, J.: Deep residual learning for image recognition. In: Proceedings of the IEEE conference on Computer Vision and Pattern Recognition (CVPR 2016), pp. 770–778 (2016)
13. Hidayatullah, P., Konik, H.: CAMshift improvement on multi-hue and multi-object tracking. In: Proceedings of the 2011 International Conference on Electrical Engineering and Informatics, pp. 1–6 (2011)
14. Krizhevsky, A., Sutskever, I., Hinton, G.E.: Imagenet classification with deep convolutional neural networks. Adv. Neural Inf. Process. Syst. (NeurIPS 2012) **25**, 1097–1105 (2012)
15. Lin, M., Chen, Q., Yan, S.: Network in network. arXiv preprint arXiv:1312.4400 (2013)
16. Liu, W., et al.: SSD: single shot multibox detector. In: Leibe, B., Matas, J., Sebe, N., Welling, M. (eds.) ECCV 2016. LNCS, vol. 9905, pp. 21–37. Springer, Cham (2016). https://doi.org/10.1007/978-3-319-46448-0_2
17. Loshchilov, I., Hutter, F.: Fixing weight decay regularization in adam (2018)
18. Lowe, D.G.: Object recognition from local scale-invariant features. In: Proceedings of the seventh IEEE International Conference on Computer Vision (ICCV), vol. 2, pp. 1150–1157. IEEE (1999)
19. Matuska, S., Hudec, R., Kamencay, P., Trnovszky, T.: A video camera road sign system of the early warning from collision with the wild animals. Civil Environ. Eng. **12**(1), 42–46 (2016)
20. Norouzzadeh, M.S., et al.: Automatically identifying, counting, and describing wild animals in camera-trap images with deep learning. Proc. Natl. Acad. Sci. **115**(25), E5716–E5725 (2018)
21. Ojala, T., Pietikäinen, M., Harwood, D.: A comparative study of texture measures with classification based on featured distributions. Pattern Recogn. **29**(1), 51–59 (1996)
22. Pinto, F., Torr, P., Dokania, P.K.: Are vision transformers always more robust than convolutional neural networks? In: Advances in Neural Information Processing Systems (NeurIPS 2021) (2021)
23. Redmon, J., Farhadi, A.: Yolo9000: better, faster, stronger. In: Proceedings of the IEEE conference on Computer Vision and Pattern Recognition (CVPR 2017), pp. 7263–7271 (2017)
24. Ren, S., He, K., Girshick, R., Sun, J.: Faster r-CNN: towards real-time object detection with region proposal networks. arXiv preprint arXiv:1506.01497 (2015)
25. Schneider, S., Taylor, G.W., Kremer, S.: Deep learning object detection methods for ecological camera trap data. In: Proceedings of 2018 15th Conference on Computer and Robot Vision (CRV), pp. 321–328. IEEE (2018)
26. Simonyan, K., Zisserman, A.: Very deep convolutional networks for large-scale image recognition. arXiv preprint arXiv:1409.1556 (2014)
27. Swanson, A., Kosmala, M., Lintott, C., Simpson, R., Smith, A., Packer, C.: Snapshot serengeti, high-frequency annotated camera trap images of 40 mammalian species in an African savanna. Scientific Data **2**(1), 1–14 (2015)
28. Swinnen, K.R., Reijniers, J., Breno, M., Leirs, H.: A novel method to reduce time investment when processing videos from camera trap studies. PLoS ONE **9**(6), e98881 (2014)
29. Szegedy, C., et al.: Going deeper with convolutions. In: Proceedings of the IEEE conference on Computer Vision and Pattern Recognition (CVPR 2015), pp. 1–9 (2015)
30. Szegedy, C., Vanhoucke, V., Ioffe, S., Shlens, J., Wojna, Z.: Rethinking the inception architecture for computer vision. In: Proceedings of the IEEE Conference on Computer Vision and Pattern Recognition (CVPR 2016), pp. 2818–2826 (2016)
31. Wu, B., et al.: Visual transformers: token-based image representation and processing for computer vision. arXiv preprint arXiv:2006.03677 (2020)
32. Wu, Y., Kirillov, A., Massa, F., Lo, W.Y., Girshick, R.: Detectron2 (2019)
33. Yu, Y., Li, Y., Quian, T.: Automatic species identification in camera-trap images. Tech. rep, Stanford InfoLab (2018)

34. Zhang, Z., He, Z., Cao, G., Cao, W.: Animal detection from highly cluttered natural scenes using spatiotemporal object region proposals and patch verification. IEEE Trans. Multimedia **18**(10), 2079–2092 (2016)
35. Zhou, D.: Real-time animal detection system for intelligent vehicles, Ph. D. thesis, University of Ottawa (2014)

Author Index

Antsiperov, Viacheslav 52
Arseneau, Lise 119

Banerjee, Anoushka 152
Bhavsar, Arnav 152
Borras, Kerstin 3

Castrillón-Santana, Modesto 134

Dinesh, Dileep Aroor 152
Ding, Steven X. 23

Eisler, Cheryl 119

Freire-Obregón, David 134

Gepperth, Alexander 76
Gunasekara, Charith 119

Hernández-Sosa, Daniel 134

Isern-González, José 134

Kershner, Vladislav 52
Krücker, Dirk 3

Lorenzo-Navarro, Javier 134

Mashechkin, Igor 98

Petrovskiy, Mikhail 98

Rehm, Florian 3
Reimann, Jan Niclas 23

Saletore, Vikram 3
Santana, Oliverio J. 134
Schak, Monika 76
Schwung, Andreas 23
Shukla, Bhargav Bharat 23

Vallecorsa, Sofia 3
Vasilev, Iulii 98

Printed in the United States
by Baker & Taylor Publisher Services